Explaining Electricity

Student Exercises and Teachers' Guide for

Grade Nine Academic Science

Jim Ross — *The University of Western Ontario*

Mike Lattner — *Algonquin and Lakeshore Catholic District School Board*

rosslattner
educational consultants — *London Ontario Canada*

Library of Congress Cataloging in Publication Data

Authors   Jim Ross
                         Mike Lattner

Printer   CreateSpace
Cover Design   Images, London, Ontario

ISBN   978-1-897007-00-6

Offices   London Ontario Canada

To teachers, parents and students everywhere who desire to bring about new ways of understanding the world.

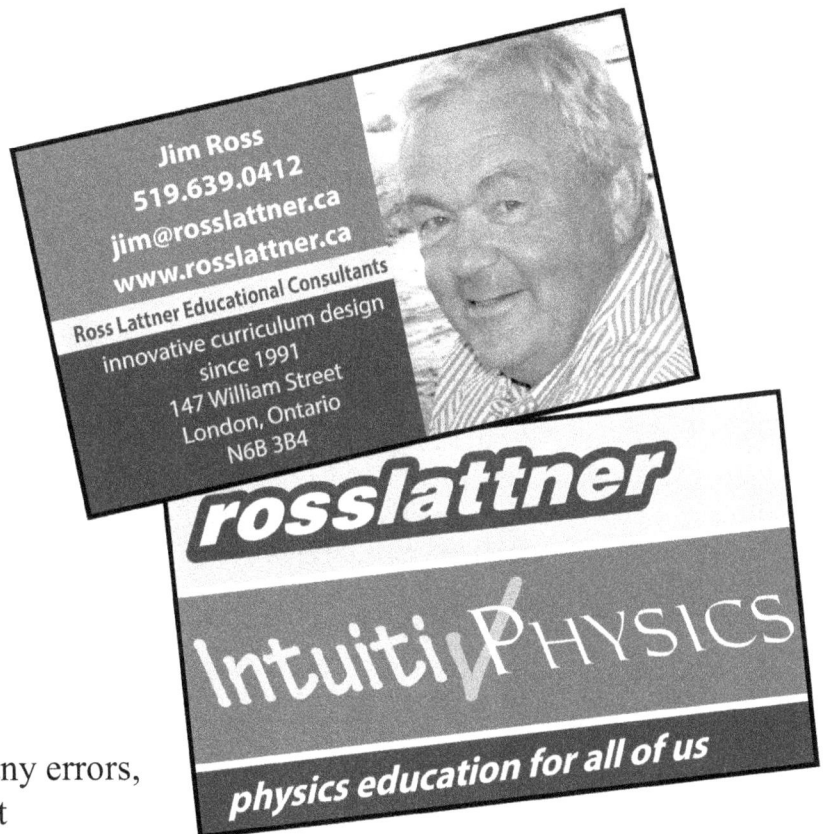

Jim Ross
519.639.0412
jim@rosslattner.ca
www.rosslattner.ca
Ross Lattner Educational Consultants
innovative curriculum design
since 1991
147 William Street
London, Ontario
N6B 3B4

rosslattner

Intuiti✓PHYSICS

physics education for all of us

We welcome your comments and suggestions.

Let us know what you find most useful.

We've worked hard to remove any errors, but don't let a day go by without letting us know if you find one.

Stay in touch.

Jim Ross

Our thanks to all of the wonderful people at the Faculty of Education, the Unversity of Western Ontario.

Explaining Electricity

Table of Contents

1: Teaching Electricity **1**

Unit Planning Notes: 2
Lab 1.1: Charging Up 4
Lab 1.2: Electrons and Charge 6
Quiz 1.3: Matter and Static Electricity 8
Lab 2.1: Getting a Light Bulb to Glow 10
Lab 2.2: What's Inside a Light Bulb? 12
Lab 2.3: What Makes Light Bulbs Different? 14
Lab 2.4: Getting Two Identical Light Bulbs to Glow 16
Lab 2.5: Getting Two Different Light Bulbs to Glow 18
Quiz 2.6: Energy and Electric Current 20
Activity 3.1: Scientists' Ideas of V, I, R & P 22
    Electric Potential 24
    Electric Current 25
    Electric Resistance 26
    Electric Power 27
Lab 3.2: Series and Parallel Circuits 28
Activity 3.3: Calculating V, I, R and P 30
Quiz 3.4: Simple Electric Circuits 32
Lab 4.1: Getting Three Light Bulbs to Light 34
Project 4.2:   Predicting and Measuring Electric Potential 36
Project 4.3:   Common Electrical Devices 38

Explaining Electricity

Table of Contents

2 Explaining Electricity **41**

Introduction: Two Theories of Electricity. 42

Lab 1.1: Charging Up 44

Lab 1.2: Electrons and Charge 46

Quiz 1.3: Matter and Static Electricity 48

Lab 2.1: Getting a Light Bulb to Glow 50

Lab 2.2: What's Inside a Light Bulb? 52

Lab 2.3: What Makes Light Bulbs Different? 54

Lab 2.4: Getting Two Identical Bulbs to Glow 56

Lab 2.5: Getting Two Different Light Bulbs to Glow. 58

Quiz 2.6:   Energy and Electric Current. 60

Activity 3.1: Scientists' Ideas of V, I, R, & P 62

    Electric Potential. 64

    Electric Current. 65

    Electric Resistance 66

    Electric Power. 67

Lab 3.2: Series and Parallel Circuits. 68

Activity 3.3: Calculating V , I , R and P 70

Quiz 3.4:   Simple Electric Circuits. 72

Lab 4.1: Getting Three Light Bulbs to Glow 76

Project 4.2: Predicting and Measuring Potential. 78

Project 4.3:   Common Electrical Devices. 80

Appendix: Laboratory Safety **82**

1: Teaching Electricity

Title:	Explaining Electricity
Time Allocation:	27.5 hours (22 periods of 75 minutes each)
Authors:	Jim Ross and Mike Lattner
Date:	May 2003
Unit Description:	An exploration of basic electrical phenomena, this unit emphasizes development of a fruitful model of simple electric circuits. Because this unit involves fundamental forces and concepts, it is placed first in the series of four strands. It should be taught before chemistry.

The unit itself is further subdivided into four sections, approximating the four weeks spent on Electricity.

1. Static electricity and the electrical structure of matter
2. Characteristics of electric current, and development of a model of current, potential, resistance and power
3. Mathematical treatment of series and parallel circuits
4. Projects that are either an application of the model or an extensions of the model.

At the end of sections 1 - 3 is a thorough quiz

Strand:	Physics
Expectations:	Overall Expectations: PHV 01 - 03 Specific Expectations: PH1.01 - .13; PH2.01 - .06

Explaining Electricity

Kids possess a natural intelligence which provides them with their sense of meaning in the world. This intelligence appears to be structurally different from the discipline of science.

We need to teach kids to use their natural ability to "make meaning," but teach them to use it in the way that a scientist might use it.

Unit Planning Notes:

This unit extends the particle theory of matter. The particles of matter are represented as being composed of light, mobile, negative electrons and a massive, stationary, positive nucleus. Electrical phenomena are explained in terms of the movement of electrons from regions of high energy to regions of low energy.

In addition, this unit emphasizes the *explanatory* and *predictive* roles of science, ahead of the empirical. The lab exercises are not designed to provided definitive, once-and-for-all answers to highly refined questions. That would place the exercises outside the realm of ordinary thought of fourteen-year-old human beings. Instead, the lab exercises are designed to provoke student prediction, and subsequent explanation, of selected phenomena.

To this end, the introduction of the usual technical language of *current*, *voltage*, and *resistance* is introduced at the last possible moment. Instead, visual representations of current flow and energy transformations are promoted in this unit. These representations have been carefully arranged to correspond to the ways that young people appear to think. Our intent is to anticipate the structures of natural thinking that kids use every day to make sense of the world, and organize the labs around these structures.

Prior Knowledge Required

This unit assumes that the student is quite familiar with applications of household AC, but is not likely to have a reliable grasp of basic AC concepts. The student is also expected to be familiar with some applications of DC, but has had limited experience with low voltage DC circuits. A knowledge of the particle theory of matter is expected. Because of its emphasis upon explanation, this unit expects that the student is capable of writing coherent sentences and paragraphs.

Teaching and Learning Strategies

Students don't learn what comes into them via their senses. Students learn what comes out of themselves via their representations.

The focus of science is not nature itself. The focus of science is our shared *representations* of nature. Accordingly, three learning strategies are emphasized. Students are expected to commit themselves to a *prediction* of the behaviour of each demonstration or lab. Students are expected to explain why they believe their prediction, in both *pictorial representations* and in *sentences*. Students are expected to gradually master a small set of *theoretical propositions*, and then to increasingly represent their arguments in terms of the theory.

Assessment and Evaluation

A variety of strategies and instruments will be used throughout this document.

Introduction

The student learning goals of this unit are:

1. to predict qualitatively the behaviour of light bulbs in a circuit

2. to measure potential, current, and resistance in a set of circuits.

Six simple statements which can be used to explain things that happen around us, and even help us to predict things we have never seen!

1. **Electrical Composition of Matter.** A student's understanding of static electricity depends upon a working notion of a *field* in a *three dimensional space*. This is beyond the capabilities of grade nine students. The study of static electricity in this unit, therefore, is limited to proposing and supporting a model of the electrical composition of matter.

   1. **Atoms are made of a massive, positive nucleus surrounded by light, negative electrons.**
   2. **Opposite charges attract. Like charges repel.**
   3. **The outer electrons in an atom can be moved around.**
   4. **The nucleus and inner electrons are too massive to move.**
   5. **Losing electrons makes matter positively charged.**
   6. **Gaining electrons makes matter negatively charged.**

2. **A Theory of Energy and Electric Current**

   1. **Electrons can gain or lose electric potential energy.**
   2. **The electrons in a single conductor have the same energy.**
   3. **Electrons move from high energy to low energy.**
   4. **Electrons must spend all of their energy before they return to the source of potential.**
   5. **Resistance reduces the flow of electrons.**
   6. **The more energetic the electrons, the greater number that get through a resistance.**

3. **A Model of Current, Potential and Resistance**
   This algebraic model is not easily accessible by students of any age. The first step must be to become as familiar as possible with the dynamics of current flow and potential difference. Attempting to learn to apply these equations without a conceptual understanding of current and potential will almost certainly result in a high level of student failure and frustration.

$$R = \frac{V}{I}$$

$$I = \frac{V}{R}$$

$$V = IR$$

$$P = VI$$

$$R_T = R_1 + R_2 + R_3 + \ldots \qquad \frac{1}{R_T} = \frac{1}{R_1} + \frac{1}{R_2} + \frac{1}{R_3} + \ldots$$

$$I_T = I_1 = I_2 = I_3 = \ldots \qquad I_T = I_1 + I_2 + I_3 + \ldots$$

$$V_T = V_1 + V_2 + V_3 + \ldots \qquad V_T = V_1 = V_2 = V_3 = \ldots$$

$$P_T = P_1 + P_2 + P_3 + \ldots \qquad P_T = P_1 + P_2 + P_3 + \ldots$$

Lab 1.1: Charging Up

Complex expectations, such as this one, cannot be met in a single step. This activity continues over two days.

Learning Expectation 1.01D: Describe the properties of static electric charges, and explain electrostatic attraction and repulsion using scientific models of atomic structure.

Pedagogical Issues

These phenomena require at least two steps to explain: charge transfer, and subsequent static electrical force. Students tend to naturally prefer one-step explanations.

Worth repeating! We are sentient beings. A sentence is a single thought. Teachers must insist that students speak and write in complete sentences if they are to construct in their own beings the culture of thinking.

By the end of the lab exercise, each prediction and explanation should be accompanied by a brief sentence that includes both an account of the charges on the objects, and of their subsequent interaction.

Kids have a well developed narrative sense, even if they are unfamiliar with the characters. Treat each situation as a little story about electrons that develops over time. Voila: instant theoretical account!!

Science Issues

Throughout this unit, student understanding of electricity is assessed as "the extent to which the student uses a theoretical representation of the structure of matter to correctly predict and explain the behaviour of simple circuits." The purpose of these three lessons is to contribute to the formation and testing of an appropriate representation of a model of matter.

You want your students to leave this unit with an appreciation for the value of a good theory.

It is most important that the representations used have the greatest generality for the future. The models proposed here appear to be useful up to the point of electron-electron interactions at the quantum level.

Static Electricity

Foil covered pith balls pick up electric charge more readily. You may use the foil from cigarette packages: heat the foil with a lighter, peel off the paper backing, and wrap pieces of the waxy foil around the ball.

These activities are best conducted on a dry day. Atmospheric humidity can greatly inhibit the ability of an object to hold charge.

The Learning Activity

There are 5 activities in this lab. Before each activity, the students should make a prediction of the behaviour. Their prediction and their explanation of the prediction should be written before they do the experiment. Following the experiment, students record in pictures and words both their observations and their new explanations.

Students generate electric charges on plastic strips, transfer them to pith balls, and then observe the nature of the interactions among the various charged strips and pith balls.

Acetate becomes quite positive when rubbed with paper; vinylite becomes somewhat negative when rubbed with fur or wool.

Alternatively, you may wish to hang the plastic strips from strings, and have them interact directly.

Equipment, Preparation and Resources

Sufficient equipment for a class of 24, working in pairs:

12 vinylite and 12 acetate strips
fur or wool, and paper
12 pith balls on threads.

Categories:	Assessment and Evaluation
Knowledge:	answers to *Questions for Later...*
Inquiry:	clarity and coherence of predictions and explanations
Communication:	clarity of diagrams and sentences in written work
Applications, Extensions:	

Explaining Electricity

Lab 1.2: Electrons and Charge

Learning Expectation 1.02D: Describe the properties of static electric charges, and explain electrostatic attraction and repulsion using scientific models of atomic structure.

Ontology is the study of being, of things in themselves, of things as they are.

The most intuitively meaningful ontological structure is that of matter. Children's ontology of novel things is frequently identical to their ontology of matter.

Pedagogical Issues

The key difficulty for students appears to be their lack of an appropriate ontological structure; students speak of *"electricity"* as if it was a thing in itself, and not a composite phenomenon. Scientists, on the other hand, speak of electricity more fruitfully in terms of *electrons*, *matter* and *energy*. Therefore, the key learning task is the consistent representation of the electron, its energy, and its relation to the rest of matter.

It is unlikely that students have formed a naive understanding of the structure of matter. It is therefore expected that the students will make simple operational predictions without coherent explanation. The purpose of this activity is to provide an ontological basis from which a coherent explanation of electricity can be made.

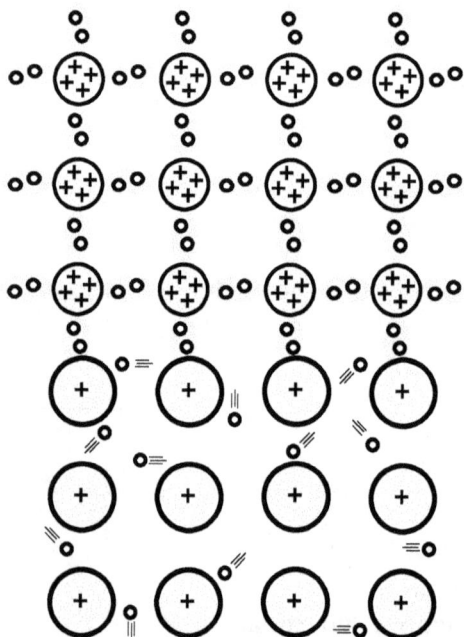

Science Issues

An appropriate representation of matter is required. If this unit precedes chemistry, the distinction between conductor and non-conductor is very closely parallel to the chemists' distinction between metal and non-metal.

Non-metals are depicted as having smaller cores with many positive charges. They have many electrons, which are tightly held. This corresponds to the chemists' notion of covalent bonds.

Metals are depicted as having larger cores with few positive charges. Its weakly held electrons are free to move from atom to atom.

Static Electricity

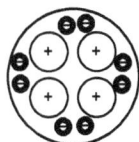

Happy electrons lounge in a metal bar. Excess electrons approach.

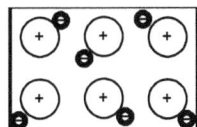

Frightened, our happy electrons flee to the end of the bar.

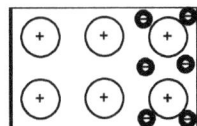

A bridge to ground appears. Our electrons flee to the ground.

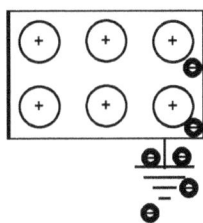

The bridge to ground breaks. The two remaining electrons cower at the far end of the bar.

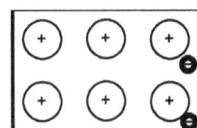

The excess electrons depart. The bar has a deficiency of electrons.

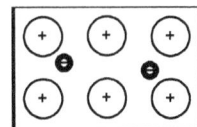

The Learning Activity

You will need to use an electrostatic machine to demonstrate the charging of other objects by conduction (contact), and by induction.

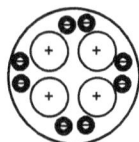

Charging by conduction is intuitively simple. Students' everyday thinking is quite appropriate in this case.

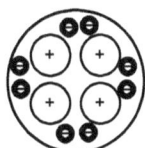

Charging by induction requires a more extended narrative structure, perhaps beyond the ability of some students. In the example at left, a charged object is brought near a metal bar, the metal bar is briefly touched with a finger, and the charged object is withdrawn. The metal bar now bears a charge opposite to the original charge.

Transmission of these images to the student in the form of a note is not very effective in the long run. Instead, this or a similar structure should gradually emerge as students struggle to explain inducted charge within a directed class discussion.

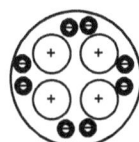

Equipment, Preparation and Resources

Sufficient equipment for a class of 24, working in pairs:

12 vinylite and 12 acetate strips
fur or wool, and paper
12 pith balls on threads.

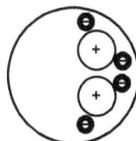

Static Electricity machine, either Whimhurst or Van de Graaff.
Several insulated metal balls
Charged pith balls or other electroscopes, used to determine whether other charged objects are positive or negative.

Categories:	Assessment and Evaluation
Knowledge:	Demonstrates coherent use of model of matter in explanations.
Inquiry:	
Communication:	Writes a clear *sequential* account of charging objects.
Applications, Extensions:	

Explaining Electricity

Quiz 1.3: Matter and Static Electricity

Learning Expectations: PH 1.01 - 1.03 Electrical composition of matter, electrostatic forces, and the charging of non-conductors by various means.

Pedagogical Issues

These exercises afford the student an opportunity to test his or her knowledge. These items are intended to provoke some degree of cognitive activity, rather than recognition or recall.

They are also designed to require a facility with the system of notation which we are developing, in which electrons are represented by open circles ○ . Consistency in notation is one of the characteristics of science.

During the historical period in which new ideas are being expressed, several notational systems may be used. It is a remarkable characteristic of science that these tentative systems are so quickly critiqued. Most of them rapidly discarded, and within a decade or two completely forgotten. Upon the notational systems which are kept, however, a new community of scientists forms. These scientists, enjoying a common form of public representation, can occasionally build a new science.

Science Issues

An attempt has been made here to build "upward compatibility" into these items. The representations are consistent with, for example, the Lewis Dot convention used to describe chemical bonding.

The Learning Activity

These quizzes can be used profitably in several ways:

Daily Pop Quiz. Did the kids do the homework? Did they understand it? You can pop one of these questions on the class the day after the lesson, and quickly assess problems.

Daily Practice Quiz If half the class could do it on Tuesday, can they improve by Thursday?

Discussion Generator Some questions and responses can generate controversy in the classroom. When students are required to explain their beliefs, some very fruitful learning situations can develop.

Question on a later summative test Feel free to use any of these quiz items on a summative test. Students respond more confidently to structures they have seen before.

Equipment, Preparation and Resources

Quizzes in the lab manual, pencil, eraser, etc.

Categories: **Assessment and Evaluation**
Knowledge:
Inquiry:
Communication:
Applications, Extensions:

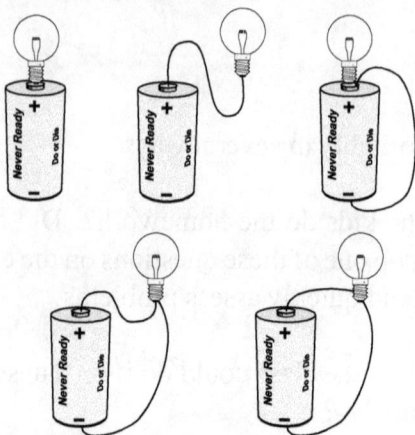

Lab 2.1: Getting a Light Bulb to Glow

Five common predictions of ways to get a bulb to light up. (From *Driver et al 1994*)

Unipolar model
Source-Path-Goal
schema

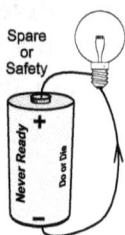

Spare
or
Safety

Clashing Current
model
Conflict schema

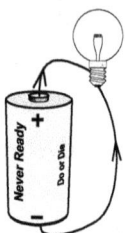

Diminished current
model
Effort schema

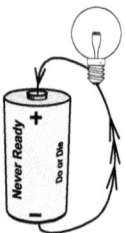

Conserved current
The scientists' model
of current

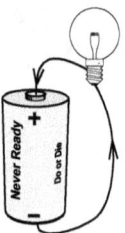

Learning Expectations 1.04B, 1.09D, 1.10D: Concepts of current, potential, and resistance; their behaviour in series and parallel circuits.

Pedagogical Issues

Research has shown that electric potential, current, and resistance are some of the most difficult ideas to grasp. Students naturally think of potential as an effort exerted by a source, rather than as potential energy. The purpose of this series of lessons is to challenge students to test and refine their intuitive model of electricity.

Students are likely to predict one of the five systems top left. They may or may not have coherent reasons for doing so. When asked to explain, however, they usually speak of something moving to the lamp, and being used up there.

In the international literature on science misconceptions, several conceptions of the operation of electrical circuits are commonly reported. Only one of these corresponds to scientists' ideas.

Each of the other models corresponds closely with one common mental schema. Students find these schematic models so convincing, however, that in the absence of profound conceptual change they will prefer to use them over the scientists' model.

It is most important, therefore, that sufficient time be devoted to cultivating the scientific model, and challenging the students' incorrect ideas.

The **Predict, Explain, Observe, Explain** cycle is one way to require students to commit their personal ideas to paper. As their ideas are represented in a public fashion, classroom discussion, experiment, and direct instruction can be organized or steered so as to challenge inadequate student ideas. At the same time, carefully directed discussion can help to increase student commitment to the scientific model, and reduce commitment to their own schematic models.

Energy and Electric Current

To *explain* is to tell *why* the bulbs light up, without using the words *bulb* and *light*.

We are *sentient* beings, *thinking* beings. We must express our thoughts in sentences.

One sentence is needed to express one thought. If you must express another thought, write another sentence.

The Learning Activity

Given a battery, a bulb, some aluminum foil, copper wire, paper and plastic, the students attempt to get the bulb to light up.

Before the experiment

Predict: the student must think about how to get the bulb to light up, and commit to a plan.

Explain: the student must commit to a line of reasoning that tells why or how the arrangement works. A little discussion at this point helps hesitant students.

After the experiment

Observe: students make representations of the arrangement that actually worked.

Explain: how the working circuit actually functions. Time (30 minutes) should be left for a classroom discussion.

Equipment, Preparation and Resources

A number of authorities recommend the use of 1.5 V dry cells in this experiment, because they are familiar and simple. On the other hand, they are costly, they wear out, and their potential declines with use.

Power supplies are more reliable, but they are unfamiliar and complicated, and may distract students from the main ideas. The maximum potential of a lab power supply is larger, and the lamps must be able to withstand at least 6 V.

One way to keep this set of labs organized is the use of kits. Twelve kits would serve a class of 24. Each kit contains:
1 cell or power supply
1 lamp, 1.5 volt (6 volt if using power supply)
wires, foil, paper, plastic, etc.

Categories: **Assessment and Evaluation**

Knowledge: The last explanation of the PEOE cycle.

Inquiry:

Communication: Clarity and quality of student drawing and explanation.

Applications, Extensions:

Lab 2.2: What's Inside a Light Bulb?

Learning Expectations 1.03B, 1.04B: Current electricity vs static electricity; concepts of electric current, potential difference, and resistance.

Pedagogical Issues In the last lab, we raised the issue of student ideas about electricity. Students cannot grasp what electricity is, and how it works, apart from an understanding of what is inside the light bulb. The function of the circuit cannot be separated from the function of the light bulb itself.

A characteristic of "everyday thinking" or "natural intelligence" is that it does not require such conceptual integrity. Students are often content to comprehend electricity and the light bulb separately, even with mutually incompatible models.

One common model of electricity that is conceptually separate from a model of the light bulb is the "clashing current model" of the circuit. The light bulb is the site of a clash between one kind of electricity from one end of the cell and a different kind of electricity from the other. This conception is typical of the *conflict* schema.

Such conceptions must be challenged. The teacher's questions should challenge the student to make a representation of the idea, in pictures and words.

"What does a clash of currents look like? *Draw me a picture!*"
"Why would the current clash inside the bulb, and not outside the bulb?

A good question requires that the student produce a real thought.

Questions which require single word answers cannot get at student understanding.

Science Issues In the diagram at left, every electron ○ leaving the negative end of the cell has six units • • • • • of energy. Electrons can travel through conductors to the light bulb without losing energy. Electrons must spend all of their energy before returning to the positive end of the battery. In this case, each electron spends all of its energy in the single resistance.

In this example, six electrons pass through the resistance, each one spending all of its energy. The total energy spent then is thirty-six units. This bulb produces a lot of light energy!

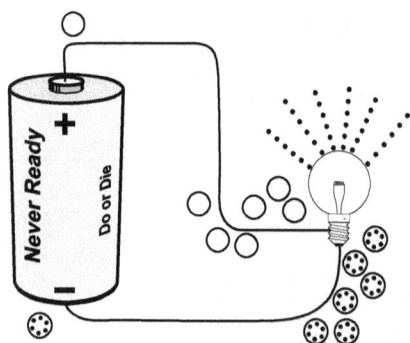

The Learning Activity

Safety is an issue here. Broken glass and sharp metal fragments must be dealt with.

Suggestions:

The value of the lab or demonstration is not found in the materials. It is found in the quality of the questions that the students ask.

Conduct this class as a demonstration

Prepare some light bulbs ahead of time for display.

Prepare some light bulbs at various stages of dissection, and have students circulate.

Have closely supervised students conduct the dissections in stages, for later display.

Equipment, Preparation and Resources

A number of light bulbs
At least 2 heavy plastic bags, eg. rinsed, dried 1 L milk bags.
1 hammer
1 file
2 pairs of needle nose pliers
1 pair of side cutters or wire snips
Magnifying equipment. Dissection microscope, large magnifier.

Safety glasses are not optional.

Use double bags!! If you put the file inside the bags with the light bulb, and hold a sharp corner of the file against the glass bulb, a light tap with a hammer on the outside of the bag will easily break the bulb.

After breaking the bulb, handle only with pliers.

What is your plan for disposing of the broken glass?

Categories:	Assessment and Evaluation
Knowledge:	correct use of the terms conductor, insulator, resistor
Inquiry:	structure of the student's inquiry
Communication:	
Applications, Extensions:	should we discard our tungsten?

Explaining Electricity

Lab 2.3: What Makes Light Bulbs Different?

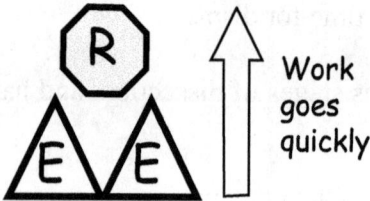

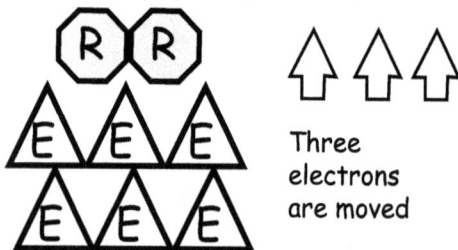

Learning Expectations 1.04B, 1.07 D: We are attempting to deepen student understanding of the relationships among resistance, current, and potential

Work goes slowly

Work goes quickly

Work goes very slowly

Three electrons are moved

Pedagogical Issues   Of all of the concepts in electricity, *resistance* is the one which has the greatest relationship to natural modes of thinking. *Resistance to effort* is an integral part of everyday experience (see left), and appears to exist as one of the fundamental schemata used in everyday thinking. This schema organizes a relationship between three things: an effort, a resistance, and some visible outcome.

If students spontaneously organize their ideas about a glowing light bulb according to the *Effort* schema, they will think of electrons as beinghe conscious, willful agents which exert effort to get through a barrier. Students are likely to suggest that electrons will "try" to get through, "exert more effort" etc.

Explicit student use of this anthropomorphic strategy should be carefully avoided. If students are "successful" at predicting and explaining things in anthropomorphic terms, they will not see the need to change their ideas to more scientific concepts. This will almost certainly lead to problems in more advanced grades.

On the other hand, the idea that an electron with 6 units of effort, acting on 2 units of resistance, provides 3 units of results is quite intuitive for students. Use what you can of students' intuition.

Science Issues   Each electron has six units of energy. That energy must be completely spent in the resistance in the circuit, in this case, the filament.   Because this filament has two units of resistance,
6 energies / 2 resistances = 3 electrons flow through.

Twice as much resistance means half as much electron flow.

Note that the total energy spent in the bulb, indicated by the "energy dots", is now also cut in half. We are not explicitly dealing with time in these representations. Each picture represents an equal amount of time, and the energy dots then represent the power of the bulb in that circuit.

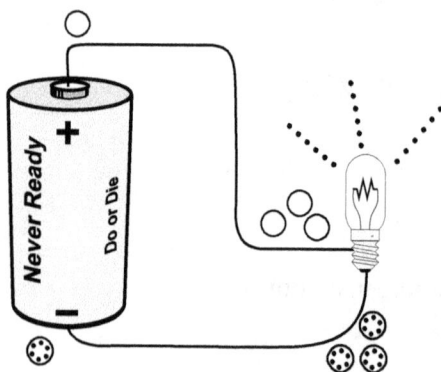

Never Ready

Do or Die

The Learning Activity

Given one socket, one dry cell, two different light bulbs, and some wire, the student task is to make each light bulb light up. One of these bulbs has more resistance than the other. *What is it about different light bulbs that makes them bright or dim?*

Before the experiment

Predict: Which bulb will be brighter, the one with the least resistance, or the one with the greatest?

Explain your prediction, using diagrams, and the words and ideas used in the previous labs.

After the experiment

Observe your experiment, and make records of your observations.

Explain your observations, using pictures and sentences.

Natural, schematic apprehension of physical phenomena is so emotionally satisfying (and therefore so persistent) that in many studies, even highly educated adults in science-related fields have been found to use schematic thinking uncritically to describe electric circuits.

Equipment, Preparation and Resources

The dry cell vs. power supply controversy continues. Dry cells are cognitively simpler, and focus the learning on the circuit. On the other hand, their usefulness declines over time. Power supplies are complex, unfamiliar, and may be distracting, but they are more reliable. In addition, a greater variety of bulbs is available for 6 V than for 1.5 V applications.

Each kit should contain:

1 cell or power supply
1 lamp socket
2 different lamps, 1.5 volt (6 volt if using power supply)
wires.

Categories:
Knowledge:
Inquiry:
Communication:
Applications, Extensions:

Assessment and Evaluation

the last explanation of the PEOE cycle

clarity and quality of student drawing and explanation

Explaining Electricity

Lab 2.4: Getting Two Identical Light Bulbs to Glow

Learning Expectations 1.04B, 1.07D, 1.09D: Current, potential, and resistance . Characteristics of series and parallel circuits.

Pedagogical Issues

A concept is the totality of all of the relationships in which that concept participates (West & Pines 1984)

Everyday words such as *electricity* already have rich fields of connections with events and things in students' lives, and so are meaningful to students. When learning new concepts, meaningful learning requires that students make connections with familiar things first, and only gradually expand the set of connections to the unfamiliar and sometimes counter-intuitive ideas of scientists.

The electrons in a single conductor have the same energy.

Electrons move from high energy to low energy.

The premature introduction of technical terms such as "series circuit" and "parallel circuit" before the differences are actually clear to the learner can be fruitless, or even counter-productive. If students have no conceptual framework into which to connect these new terms, the concept "parallel circuit" remains a cognitive orphan.

Electrons must spend all of their energy in resistances before returning to the source of potential .

The notational system which we have been using allows students to work through the characteristics of series and parallel circuits without ever using the terms. If the teacher can guide students to gradually make meaningful connections with existing notions, learning is more likely to be permanent. In other words, the student can accomplish a lasting conceptual change.

Resistance reduces the flow of electrons

Science Issues

The more energetic the electrons, the greater number that get through a resistance.

If students can represent electric circuits using just three fundamentals (electrons, energy, and resistance), the classical formulations known as Kirchoff's Rules can be readily understood by students. Kirchoff's rules without Kirchoff! This permits students to make decisions about circuits without using the often problematic words "series" and "parallel." These can be introduced after students have had success at using the underlying principles.

The Learning Activity

Given two identical light bulbs, two sockets, wire, and a power supply or dry cell, students must find two distinct ways to arrange the bulbs so that both bulbs light up.

Predict how to make both bulbs light up, two different ways.
Explain how your plan will work.
    Experiment with different arrangements until they work.
Observe and record pictures of your working models.
Explain: Are the bulbs the same brightness? Why or why not?

The experiment itself may only take a few minutes, but the discussion will take a considerable amount of time.

In the first circuit, each electron must pass through one unit of resistance. Since the electron can spend all six units of energy in one unit of resistance, many electrons can get through (six). A large number of electrons spending their energy in a single bulb means lots of energy gets turned into light in each bulb.

In the second circuit, fewer electrons (three) can get through the double resistance. When they do, they cannot spend all of their energy in one bulb.

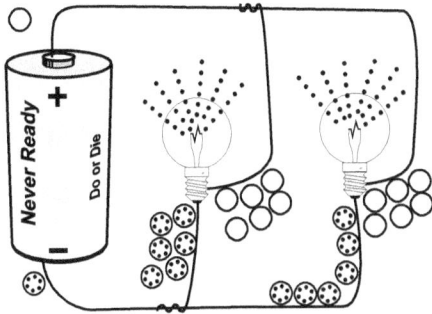

Each electron must pass through only one resistance, so 6 electrons can get through. Look at the energy!

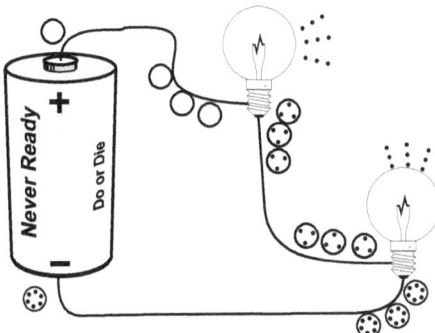

Every electron has 6 units of energy to spend. Every electron must go through 2 units of resistance (2 bulbs). The flow of electrons is reduced to 3 (6 energy÷2 resistance). Each electron spends half its energy in the first bulb, and half in the second.

Equipment, Preparation and Resources

The argument here is beginning to get quantitative. Six units of energy per electron is beginning to look a lot like six volts. Three units of charge per unit time is beginning to look like a current of three amps.

Each kit should contain:

1 cell or power supply
1 lamp socket
2 identical lamps, 1.5 volt (6 volt if using power supply)
wires.

Categories:
Knowledge:
Inquiry:
Communication:
Applications, Extensions:

Assessment and Evaluation

The last explanation of the PEOE cycle

Clarity and quality of student drawing and explanation

With six units of energy per electron, and three units of resistance, only two electrons can get through (6/3 = 2).

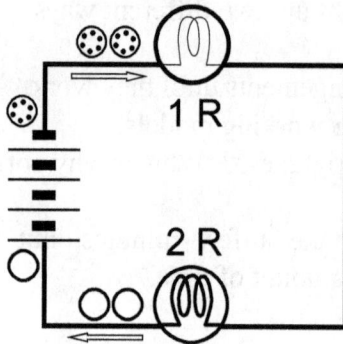

Each electron will spend more of its energy getting through the greater resistance. If an electron spends 2 energies in 1 resistance, it will spend 4 energies in 2 resistances.

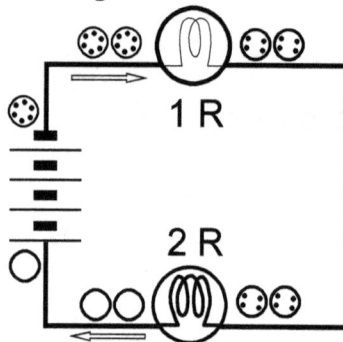

Energy spent corresponds to the brightness of the bulbs.

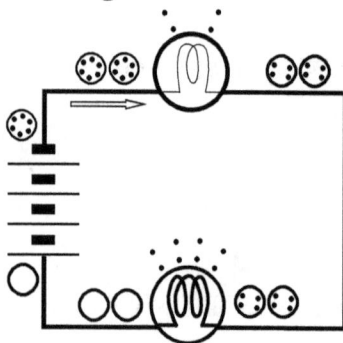

Lab 2.5: Getting Two Different Light Bulbs to Glow

Learning Expectations 1.04B, 1.07D, 1.09D: Current, potential, and resistance in series and parallel circuits.

Science Issues The sheer number of representations needed to clarify these issues suggests that this is an appropriate time to introduce schematic circuit diagrams. Steps to solve a diagram:

1. What is the total resistance that an electron must pass through in the circuit?
2. Given the total energy available to the electron, how much current will pass through the total resistance?
3. How much of its total energy will each electron spend in each resistance?
4. Check to see if every bulb is consistent.

See the sidebars for examples.

Pedagogical Issues One of the most persistently difficult ideas for people to grasp is the idea that the potential difference across two resistors in series is divided proportionally between the two resistors. Students appear to reason about circuits as a sequence of distinct events, rather than as a network of inter-related phenomena. These two misconceptions appear to be related to schematic modes of everyday thinking.

Given a circuit with two bulbs in series, many students use the *Effort* schema to predict that the first bulb will be brighter. They appear to reason that the electricity has more energy at the first bulb, and is tired out at the second bulb. The *Effort* schema, which cognitively organizes so many phenomena in everyday life, is unable to deal with the idea that the downstream bulb actually affects the upstream bulb. This lab provides an opportunity for the teacher to directly contradict that everyday idea. In the series circuit, the more resistive bulb is the brightest, no matter what its position.

Another common misconception is that the bulb brightness depends only upon the resistance of the bulb. There are bright bulbs and dim ones. This lab directly contradicts that everyday idea. The more resistive bulb is brightest in the series circuit, but dimmest in the parallel circuit.

The electrons will spend six units of energy in either bulb.

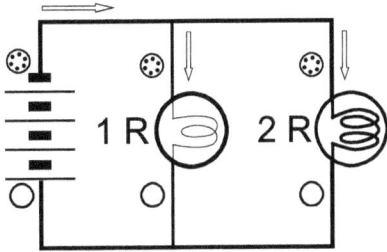

With six energies, six electrons can get through one resistance; but only three electrons can get through two resistances.

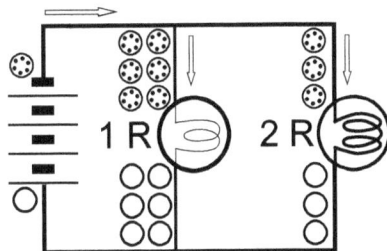

The energy spent in the bulbs corresponds to the brightness of the bulbs. Note that the more resistive bulb is dimmer in this circuit.

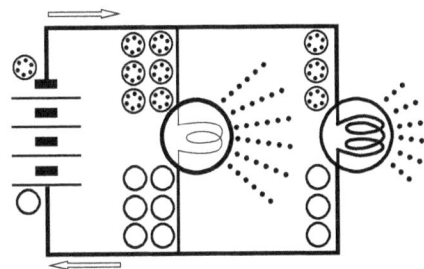

The Learning Activity Before starting the PEOE cycle, demonstrate the behaviour of the two different bulbs *by themselves* with the source of potential your class is using.

Given two different light bulbs, two sockets, wire, and a power supply or dry cell, find two distinct ways to arrange the bulbs so that both bulbs light up.

Before the experiment
   Predict: Which bulb will be brighter in every circuit you design?
   Explain in pictures and related words exactly why your bulbs will behave that way.
After the experiment
   Observe and record diagrams of your working models.
   Explain: Is each bulb the same brightness in every arrangement? Why or why not?

The experiment itself may only take a few minutes, but the discussion will take a considerable amount of time.

Equipment, Preparation and Resources

The trend toward a need for quantitative treatment continues. The relationship between energy, resistance, and flow is becoming clearer.

Each kit should contain:

1 cell or power supply
1 lamp socket
2 different lamps, 1.5 volt (6 volt if using power supply)
wires.

Categories:
   Knowledge:
   Inquiry:
   Communication:
Applications, Extensions:

Assessment and Evaluation

the last explanation of the PEOE cycle

clarity and quality of the student's drawing and explanation

Explaining Electricity

Quiz 2.6: Energy and Electric Current

Learning Expectations: Not specified

Pedagogical Issues

In the next section of this program, students will begin to deal with the more conventional treatment of current, potential, resistance, and power. The section just finished has been intended to encourage students to think coherently about the concepts which are necessary to support the conventional understanding

This quiz, therefore, is designed to test student mastery of the notational system which is being developed in the lab manual.

Science Issues

The conventional view of electric current, summed up by Ohm's Law and Kirchoff's Laws, is not at all self-evident to most human minds. A great deal of preparation is needed before such "laws" are comprehensible.

It is highly unlikely that Ohm and Kirchoff performed their investigations in the absence of a personal explanatory model of electricity. The "laws" which bear their names would likely have been incomprehensible to them without such an explanatory model, or theory.

In fact, that may be one of the differences between "law" and "theory." Can we expect students to master "laws" without at first possessing some kind of personal explanatory model, plus a notational system to allow them to express their thinking?

The Learning Activity

These quizzes can be used profitably in several ways:

Daily Pop Quiz. Did the kids do the homework? Did they understand it? You can pop one of these questions on the class the day after the lesson, and quickly assess problems.

Daily Practice Quiz If half the class could do it on Tuesday, can they improve by Thursday?

Discussion Generator Some questions and responses can generate controversy in the classroom. When students are required to explain their beliefs, some very fruitful learning situations can develop

Question on a later summative test Feel free to use any of these quiz items on a summative test. Students respond more confidently to structures they have seen before.

Equipment, Preparation and Resources

Quizzes in lab manuals, pencils, erasers, etc.

Categories: **Assessment and Evaluation**
Knowledge: correct completion of knowledge items
Inquiry:
Communication:
Applications, Extensions:

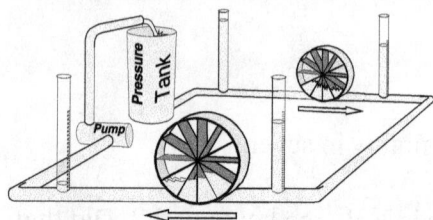

Activity 3.1: Scientists' Ideas of V, I, R & P

Learning Expectations PH1.06, 1.07: SI units of potential difference, electric current, and electrical resistance; conceptual and algebraic relationships among them.

The strength of the water analogy is its simplicity. Most students have experience with flowing water, although their ideas of pressure are full of contradictions.

Water flow is analogous to *electric current*. A movement of water at one place can only take place if the whole circuit of water moves. *Restrictions* are analogous to *resistance*.

Some difficulties arise with the idea of *potential*. In a water circuit, students think of *pressure* as the ultimate cause. *Pressure* and *potential* are thus frequently linked in the student's mind. This analogy works well in simple situations, but is not enough.

By linking potential to the analogous idea of *height* of water, the analogy is more realistic. Potential difference can thus develop whenever a resistance impedes the flow of current.

Pedagogical Issues The concepts in this cycle are among the most difficult elementary concepts in the field of science, for a number of reasons. These concepts are:

1. counter-intuitive: potential, current, and resistance do not correspond to everyday intuitive schematic thinking.

2. abstract: for example, scientists ultimately define the volt in terms of joules and coulombs, neither of which are part of everyday experience.

3. networked: a complete understanding of potential is not possible without an understanding of current.

4. confounded by everyday understandings: the word voltage is used, often incorrectly, in many everyday contexts.

One way to cut through the problems is to use an extended analogy. Two analogies are presented here: *Electricity is like flowing water*, and *electricity is like a crowd at a checkout counter*. No one analogy is perfect, or infinitely extensible. The analogies chosen here are part of everyday experience, and have many points of connection with the desired concepts *potential*, *current*, *resistance*, and *power*.

An analogy is not an end in itself, it must be used as an aid to thinking. If students have trouble thinking through a given circuit using scientists' ideas, they may be more likely to successfully predict circuit behaviour using one of these analogies.

Science Issues It is most important that the teacher not use these analogies indiscriminately. The ideas of potential, resistance, and energy are quite subtle, and require constant attention to detail.

Energy and Electric Current

The success of the money analogy is the richness and certainty of students' everyday experience with money.

The movement of people is analogous to electrical current. Money is analogous to electric potential. Power is analogous to the overall rate of expenditure of money.

The weakness of this model is the problem of resistance. There is no physical resistance to the flow of money, so an artificial restriction must be introduced. In these exercises, resistance is simply a limited availability of the desired items. Shoppers go past the checkout only when a ticket becomes available to purchase.

The Learning Activity A small group activity, related to the jigsaw structure used in cooperative learning lessons. Each student is given four groups of words, related to the four main concepts in this lesson: *potential*, *current*, *resistance*, and *power*.

Think: Students write their own ideas about each group of words on the back of their lab sheet.(5 minutes)

Share: Students get into home groups of four students (six groups of four in a class of twenty four).   They share their ideas on the four concepts. The identify ideas about wyich they are not certain. (10 minutes)

Study: One person from every home group goes to one of the four expert groups. Each expert group deals more intensely with one of potential, current, resistance and power. Each student is to take paper along, and to learn about their topic.(30 minutes)

Share: Students return to their home groups to share their ideas. (30 minutes)

The times suggested here will completely occupy a 75 minute period. Perhaps preliminary thinking can be done the night before. It may be the case that the *Study* session can be reduced in time.

Equipment, Preparation and Resources

Following this lesson are four pages in this teacher's guide. *You must make copies for the Expert Group Study session. For a class of twenty four, make six copies of each page.*

The Home Groups must have at least one individual who can help others to integrate the material.

Expert Groups must spend time discussing the analogies. This discussion time provides the largest block of time for each student to become knowledgeable about one of the four main concepts.

The two analogies are used for all four concepts, so that student familiarity with the analogy in his or her Expert Group learning will be of use in the Home Group sharing at the end of the class.

Categories: **Assessment and Evaluation**

Knowledge: application of ideas in later exercises

Inquiry:

Communication: summary of learning at the end of the lesson

Applications, Extensions:

Potential is the ability to do work

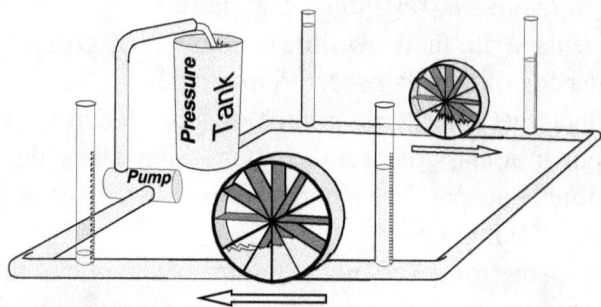

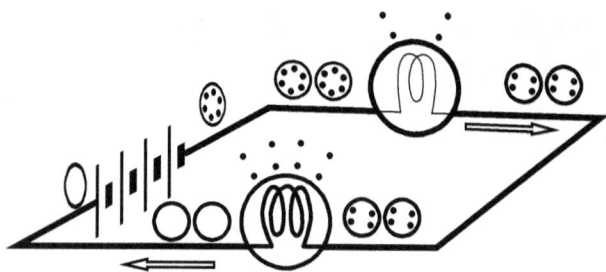

Electric Potential

Ideas

Electric potential is like water level: the higher the level, the greater the potential. A pump pushes on the water, lifting it to a high level. The water in the pipe is then under pressure. You can see the water level in the four vertical tubes. At the first water wheel, the water converts some of its potential into the mechanical energy of the turning wheel. The water level (and the pressure) are decreased. At the second water wheel, the water spends its remaining potential until the pressure reaches zero. The water must return to the pump before it can do more work.

Electric potential is like money: the more money in your pocket, the greater your potential. When you reach the first checkout, you spend some of your potential to buy a ticket. At the second checkout, you spend your money until your pockets are empty, then you go back to the bank for more.

Electric potential is *energy per electron*. A dry cell gives each electron (the circles) a small amount of energy (the dots). At the first light bulb, the electrons convert some of their energy into other forms. At the second light bulb, the electrons spend their energy until they reach a potential of zero. They can do no more work until they return to the dry cell for more energy.

Have you ever held your thumb over the end of a hose? When you press hard with your thumb, you create a lot of resistance. Do you feel how the pressure of the water increases? The water has to spend more of its potential to get past your thumb.

The equation **V = IR** means that more potential energy is spent when a lot of electric current passes through a lot of resistance.

Metric System of Units
One *Volt* equals one *Joule per Coulomb*. A coulomb is a very large number of electrons, about two million million million. Six volts means that every single coulomb has six joules of energy.

Mathematical Equations

$$V = IR$$

Electric Current is the flow of electrons

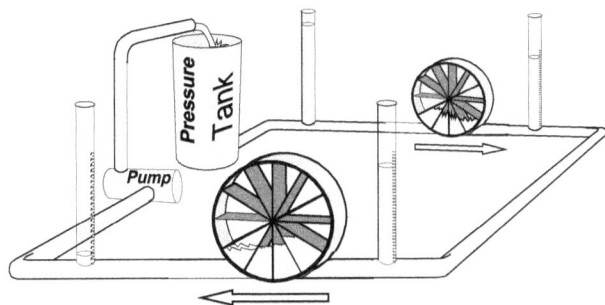

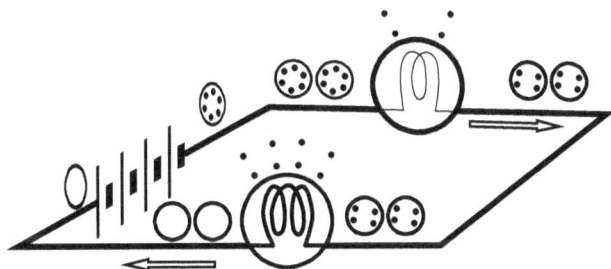

Have you ever dug a ditch in a sandbox to drain a puddle? The ditch must flow from high to low. The greater the potential, the greater the flow of water. If you plop some sand in the ditch, the water slows down and backs up. Perhaps it backs up enough to go around the resistance!

The equation **I = V/R** means that increasing the potential difference **V** causes more current **I** to flow. More resistance **R** causes less current to flow.

Electric Current

Ideas

Electric current is like flowing water: when something supplies a difference in height, the water must flow from higher potential to lower potential. All of the water flowing into a pipe or water wheel must come out. Water wheels slow the flow of water, and take energy. The resistance of the water wheel slows the flow of the water, and causes the water to "pile up" behind the resistance.

Electric current is like a crowd of people moving through a shopping mall. No one moves unless they are given some money. The more money they get, the quicker they get going! People flow from rich to poor. Everyone who goes into an aisle or a shop comes out again. Checkout counters slow the people, and take their money. The checkout resistance causes people to "pile up" into lines. If the line gets very long, people are willing to spend more money to get through.

Electric current is a flow of electrons in a circuit. If there is a potential difference, the electrons will begin to move from high to low potential through the conductors. All electrons that enter a resistor or conductor must come out; electrons cannot be created or destroyed. Light bulbs slow the flow of electrons, and convert their energy to heat and light. The resistance of the light bulbs causes electrons to "pile up," increasing the potential at that point. The more potential they have, the faster the flow of electrons through the resistance.

Metric System of Units: One *Ampere* is a flow of *one coulomb per second*, or about two million million million electrons per second.

Mathematics

$$I = \frac{V}{R}$$

Resistance costs flowing electrons energy

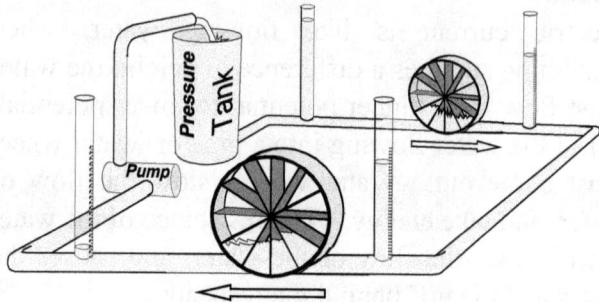

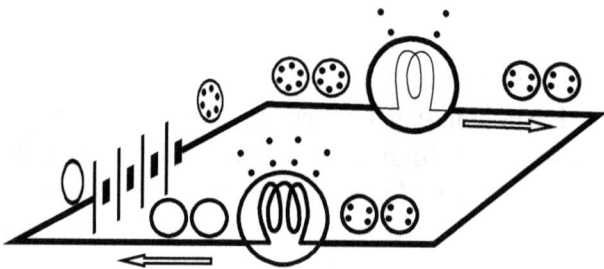

Two people let a rope slide through their hands as they lower a bucket. That's like two resistances to the motion of the rope. If one person reduces his resistance, then more rope, and more energy, is spent on the other person's hand.

The equation **R = V/I** means that if we want to know how big a resistance is, we put a potential difference **V** across it, and then observe how much current **I** flows. Small **I** means great resistance.

Electric Resistance

Ideas

Resistance is like a water wheel. Water must spend some energy to get through the wheel. The resistance turns the potential energy of height into other forms of energy. When the water comes out of the resistance, some of its potential is gone. When two water wheels are placed as shown, the same water must flow through both. Whatever potential it does not spend in the first resistance, it must spend in the second.

Resistance is like checkout counters: they slow you down, and take your money. These checkout counters are odd. They behave more like ticket outlets: everyone gets a chance to get a ticket, but only a few tickets are sold each minute. If you are allowed to buy a ticket, you still have to pay. You can imagine how much more you would be willing to pay if your chances of getting a ticket were slim, and how much less you would pay if tickets were easy to get. In this picture, the shoppers are trying to by two different tickets.

Resistance causes electrons to bumble, bounce and stick to atoms as they go through. The random bumbling and bouncing causes the atoms in the resistance to move more quickly, that is, to heat up. Resistance converts electrical potential into heat energy! The greater the resistance, the fewer electrons go through. The amount of energy actually converted depends upon the *whole* circuit.

Metric System of Units:
One *Ohm* is the resistance that will allow one *Ampere* of current to flow when one *Volt* of potential is applied.

Mathematics

$$R = \frac{V}{I}$$

Power is the rate at which work is done

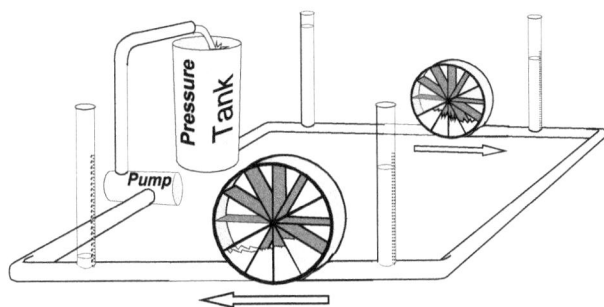

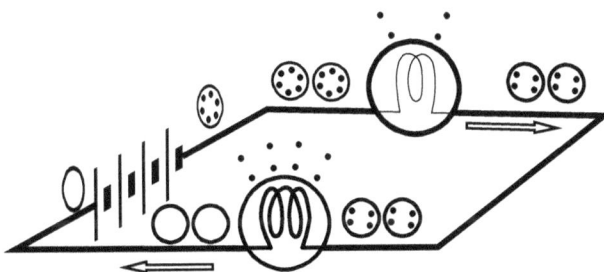

The equation *P = VI* means that power depends upon both the voltage and the current. When you turn up the dimmer switch on a 100 W light bulb, you increase the potential (volts). That increases the current (amps). More electrons go through the bulb, and each electron spends more energy. The bulb gets much hotter and brighter.

Electric Power

Ideas

The rate at which work is done in each water wheel depends upon two things. The greater the potential difference, the more work is done by every drop of water. The greater the flow of water, the faster each drop does its work on the water wheel.

Power is like the total amount of money spent on tickets in a given day. The rate at which money is spent in each checkout counter depends upon two things. The greater the price of each ticket, the more money is spent. At the same time, the more people who buy tickets, the faster the money is spent.

Power is the rate at which electrons spend their energy in a resistance. Power depends upon two things. The greater the potential difference across a resistance, the greater the energy converted to heat and light. The more electrons passing through a resistance, the faster the energy is converted.

Metric System of Units: The *Watt* is the rate of converting energy from one form to another. One Watt is a rate of conversion of *one Joule per second*.

Mathematics

$$P = VI$$

$$P = I^2 R$$

$$P = \frac{V^2}{R}$$

Explaining Electricity

Lab 3.2: Series and Parallel Circuits

Learning Expectations: PH 1.09; 1.10 describes potential difference, current and resistance characteristics of series and parallel circuits.

Meaningful learning is often described as "the student making connections between unfamiliar concepts and his or her existing knowledge."

The classroom provides an opportunity to engage the student in dialogue about the connections that the student believes are meaningful.

Perhaps the teacher's task is not simply to present "the right set of propositions" as to explore, challenge and support the student's connections as they construct their new meanings.

Pedagogical Issues   The words "series" and "parallel" have meanings in students' everyday understandings which can generally be described:

Series: things arranged sequentially, one after the other. First one thing, followed in time or space by another. One process followed by another.

This idea can be readily extended to representations of electrical components connected "in series," with the following important exception. In everyday understandings, the first process affects the second, but not the other way around. In electrical series circuits, all components affect the operation of all of the other components. Students are likely to overlook this most important characteristic of electrical circuits.

Parallel: two lines in the same direction. Two things simultaneously moving along trajectories in the same direction.

Two additional misconceptions arise quite naturally here. First, students will not "see" electrical parallelism if it is not reflected in the physical parallelism of either the physical arrangement of things, or their pencil-and-paper representation. Draw the circuits so that they *look* parallel! Second, *parallel and simultaneous* often mean *identical* to students.

Look for other cognitive connections that students may make. In logic, for example,

Series means ***both / and***: the electrons must pass through both A and B

Parallel means ***either / or***: the electrons must pass through either A, or B, but not both.

Series and Parallel Circuits

A   B   C   D

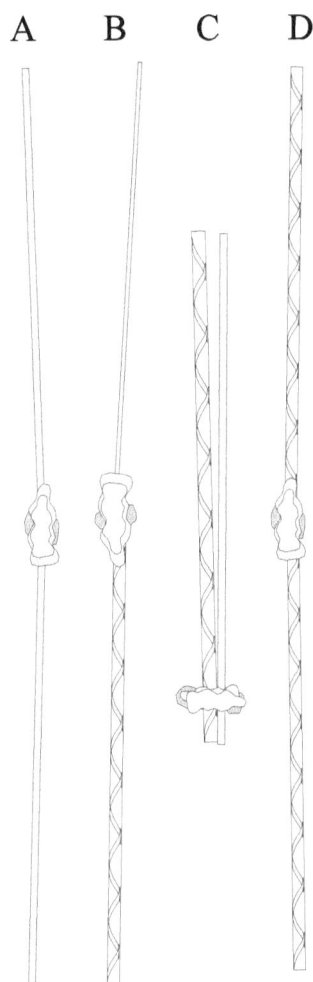

The Learning Activity

Using drinking straws, stir sticks, and some chewing gum to stick them together, students will arrange "resistances" in series and in parallel. The diagram at right provides just four of many different ways of arranging straws plus stir sticks. The number of permutations increases again if you change the direction in which students can blow through the straws. For example:

A   two equal diameter stir sticks in series
B   unequal diameter straw and stir stick in series
C   unequal diameter straw and stir stick in parallel
D   equal diameter straws in series

Before the experiment
   Predict: the air flow in each arrangement, relative to the stir stick alone
   Explain: the prediction, using previous experience or a particle diagram
After the experiment
   Observe: the effect, and make records.
   Explain: any differences from the prediction.

A very simple experiment, but capable of considerable extension.

Equipment, Preparation and Resources

6 stir sticks and two straws for each student
1 stick of gum for each student

Categories:	Assessment and Evaluation
Knowledge:	Identifies two components as either *series* or *parallel*
Inquiry:	
Communication:	
Applications, Extensions:	

Explaining Electricity

A famous metaphor in linguistic studies is the "conduit metaphor." Speech is understood as a material capsule, intentionally directed to a hearer via a trajectory, and finally "opened" and "understood" by the hearer.

This systematic, everyday metaphorical understanding of language is the cognitive basis for *hundreds* of everyday understandings of language. "I can't get through to him." "She seems a little distant." "Don't you get it?" "Do you catch my meaning?"

There is a consensus among linguists that language does not work that way. Meaning is not transmitted as little packages called words. Instead, listeners appear to construct meaning within themselves, from their own treasure of experience.

(Continued ....)

Activity 3.3: Calculating V, I, R and P

Learning Expectations: PH 1.07 - 1.10 Use the relationship $V=IR$, and apply it to both series and parallel circuits.

Pedagogical Issues We are finally here! This is the point at which many "traditional" introductory courses in electricity begin. The purpose of the past two weeks of work has been to provide the student with opportunities to construct a personal, intuitive feel for the behaviour of electrical circuits. These algebraic representations should be the completion or refinement of an understanding *which is already present*, not as the understanding itself.

Science Issues The first student exercises is:

The V_T, I_T, R_T and P_T refer to the potential, current and power provided by the energy source. These must be equal to the totals consumed in the components.

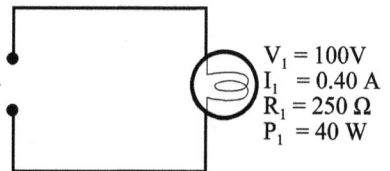

$V_T = 100$ V
$I_T = 0.40$ A
$R_T = 250$ Ω
$P_T = 40$ W

$V_1 = 100$V
$I_1 = 0.40$ A
$R_1 = 250$ Ω
$P_1 = 40$ W

The second example:

The V, I, R, and P in the components are combined as in the equations to come to the totals indicated.

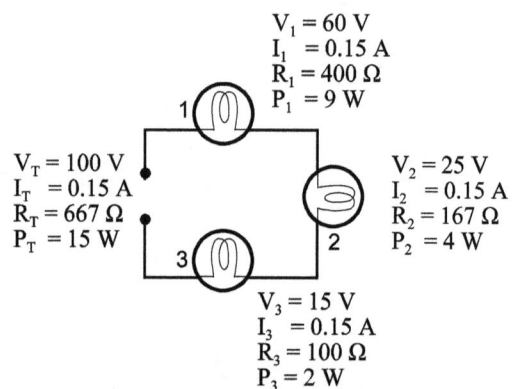

$V_T = 100$ V
$I_T = 0.15$ A
$R_T = 667$ Ω
$P_T = 15$ W

$V_1 = 60$ V
$I_1 = 0.15$ A
$R_1 = 400$ Ω
$P_1 = 9$ W

$V_2 = 25$ V
$I_2 = 0.15$ A
$R_2 = 167$ Ω
$P_2 = 4$ W

$V_3 = 15$ V
$I_3 = 0.15$ A
$R_3 = 100$ Ω
$P_3 = 2$ W

Series and Parallel Circuits

(.... continued)
In the classroom, many teaching practices simply assume that words and equations contain the intended meaning. This everyday metaphorical understanding of language has become the basis of the transmissivist view of teaching:

All that students need to do is *open the capsule*. All that teachers need is *a clear shot*.

This view is clearly inadequate.

Instead, we need to recognize that the process of *making meaning* out of traditional representations of science concepts is a very personal task.

The Learning Activity

This is a pencil and paper exercise for the students.

For the teacher, the quantities 400 Ω, 250 Ω, 167 Ω and 100 Ω correspond to the resistances of standard 25 W, 40 W, 60 W and 100 W light bulbs respectively, assuming household electricity is supplied at about 100 V.

With these numbers, you can connect the appropriate standard light bulbs into the circuits indicated, to provide a demonstration of each circuit as the students learn to solve the algebra.

Before the experiment
  Predict: the relative brightnesses, using algebra
  Explain: your prediction

After the experiment
  Observe: the actual brightness of each bulb.
  Explain: any differences between prediction and observation.

With several bulbs in each standard power rating, you could provide a considerable number of possible circuits to solve for practice.

Equipment, Preparation and Resources

At least one each of 25 W, 40 W, 60 W and 100 W standard light bulbs. The clear ones provide the best view of the filament for the very dim series circuits.

Light bulb demonstration kit, consisting of standard light bulb sockets which can be configured into appropriate circuits for demonstrations.

Categories:
Knowledge:
Inquiry:
Communication:
Applications, Extensions:

Assessment and Evaluation

correct solutions to the problems
construction and testing of other circuits.

Quiz 3.4: Simple Electric Circuits

Learning Expectations: PH1.06 - 1.10 demonstrate ability to distinguish potential, current, resistance and power; to predict all of these in given series and parallel circuits.

Pedagogical Issues

The questions on this quiz probe student reasoning about the concepts raised in this unit. Questions:

1 - 4 deal with student understanding of the analogies we used to think about electric circuits.

5 - 8 deal with issues of representation of circuits. Can students recognize conventional representations of electrical components? Can they construct series and parallel circuits?

9 - 12 deal with simple circuits containing only one component. Assessment of student mastery of the basic algebraic representations $V = IR$ and $P = VI$.

13 - 16 address series and parallel circuits of three components each. Three components form a much more critical test of student understanding than two.

Science Issues

You can treat each question like an investigation if you construct circuits to demonstrate each question item.

The Learning Activity

These quizzes can be used profitably in several ways:

Daily Pop Quiz: Did the kids do the homework? Did they understand it? You can pop one of these questions on the class the day after the lesson, and quickly assess problems.

Daily Practice Quiz: If half the class could do it on Tuesday, can they improve by Thursday?

Discussion Generator: Some questions and responses can generate controversy in the classroom. When students are required to explain their beliefs, some very fruitful learning situations can develop

Question on a later summative test: Use any of these quiz items on a summative test. Students respond more confidently to structures they have seen before.

Equipment, Preparation and Resources

Quizzes in the lab manuals, pencils etc.

Categories: **Assessment and Evaluation**
Knowledge: These are primarily knowledge, recognition items.
Inquiry:
Communication:
Applications, Extensions:

Lab 4.1: Getting Three Light Bulbs to Light

Learning Expectations: PH2.1.1 - 2.1.3; 2.2; 2.3
Conduct an inquiry using concepts in this unit, involving the construction and testing of series and parallel circuits, and qualitative assessment of their operation.

Pedagogical Issues

This lab tests the ability of students to generate new circuits, based upon previous knowledge.

Science Issues

If identical light bulbs are used, students can construct 4 different circuits. If you provide three large bulbs and three small bulbs, the number of distinctly different circuits increases to 20 different possible 3-bulb circuits.

Series and Parallel Circuits

The Learning Activity

Working in small groups, students will build different circuits to get three light bulbs to light.

Before the experiment
    Predict: as many different arrangements as you can imagine. Predict the relative brightness of each bulb.
    Explain: how the circuit should work

After the experiment
    Observe: which arrangements work, and how bright the bulbs are.
    Explain: why the arrangements work as they do.

It is vitally important that students make representations of every circuit they attempt.

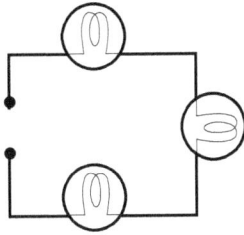

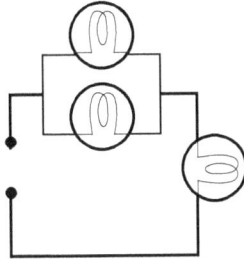

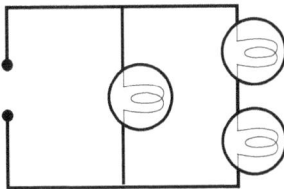

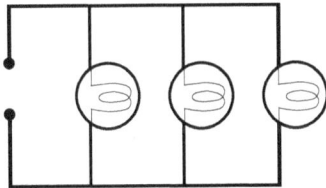

Four different circuits are possible using three identical light bulbs.

Equipment, Preparation and Resources

36 large bulbs
36 small bulbs
36 sockets
wires
12 power supplies

Categories:
Knowledge:
Inquiry:
Communication:
Applications, Extensions:

Assessment and Evaluation

quality of the students' focus questions, records, and explanation

Voltmeters must be connected *in parallel* to the component.

Think of it this way: the voltmeter samples a few electrons on either side of the component, measures their potential difference, and then puts them back into the circuit.

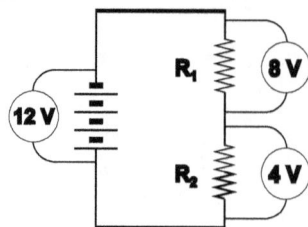

This measurement is easy, and relatively safe. If the student has set the meter to measure the highest potential provided by the energy source, then no measurement can exceed the capacity of the meter.

Following this procedure, students can safely perform this measurement without damaging the voltmeter.

Project 4.2:   Predicting and Measuring Electric Potential

Learning Expectations: PH2.1.1 - 2.1.3; 2.2; 2.3

Conduct an inquiry using concepts in this unit, involving the construction and testing of series and parallel circuits, and quantitative measurement of their operation.

Pedagogical Issues

Students are applying concepts covered and practiced in earlier labs and quizzes. Most of them should find this project moderately difficult, in the sense that applying one's knowledge to even slightly different situations requires considerable thinking, processing and doubt.

The new skill is measurement using voltmeters and ammeters. Voltmeters are simple to use, and give reliable results. Ammeters are more difficult to use, and can be destroyed in seconds. Using ammeters makes conscientious students anxious.

Science Issues

It would be wonderful to be able to use light bulbs for this experiment as well as all of the others. Unfortunately, the resistance of a light bulb changes rather dramatically with temperature. The hotter and brighter the bulb, the greater its resistance. The same bulb would have very different resistance, depending upon its placement in a circuit. Student predictions would become impossible.

Since wire wound or carbon resistors maintain their temperatures over a broader range of operating conditions, they also maintain their resistance. This allows students to make much better predictions.

Series and Parallel Circuits

An ammeter must count every single electron that passes through a component. Therefore, it must always be placed *in series* with the component. This is a more difficult operation: the circuit must be opened and the ammeter inserted before a measurement can be made. In addition, if there is no load in series with the ammeter, then all of the electrons' energy will be spent in the delicate ammeter. Thus, the ammeter is very vulnerable to instant destruction.

The Learning Activity Over a period of five days, students design four circuits, build them, predict their behaviour and test them

Day 1   Students design four circuits using three resistors. There should be about 60 different ways of arranging three different resistors.

Day 2   Calculate V, I, R, and P for each component in each circuit. Some of these will be challenging.

Day 3   Learn how voltmeters and ammeters are to be used, and then redraw the circuits so that they contain the meters.

Day 4   Measure the potential using voltmeters, ask teacher to measure current.

Day 5   Write a full report.

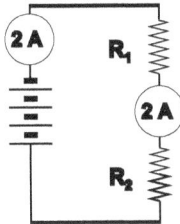

We estimate that on a typical school day in Ontario, about ten thousand dollars worth of ammeters go up in smoke.

Equipment, Preparation and Resources

For 24 students, working in pairs,

36 resistors (144 if you do not wish students to disassemble each circuit between tests)
wire
12 voltmeters, set on appropriate voltage range.

Categories:
Knowledge:
Inquiry:
Communication:
Applications, Extensions:

Assessment and Evaluation

student use of PEOE cycle, design and execution of project

3-prong grounding system
Barbecue spark lighter
Bedroom light & switch
Blow-dryer and controls
Camping lantern
Circuit breaker
Dimmer switch
Fuel Cell
Furnace Thermostat
Fuse
Ground fault interrupt GFI
Halogen Light
High pressure sodium Light
Kettle and safety shut-off
Kitchen appliance receptacles
Koolatron
Lead Acid Storage Battery
Lithium Cell
Mercury Light
Neon Light
One-cup coffee heater
Photo voltaic cell
Toaster and controls
Top / Bottom light switch
Tri-light bulb
Wind Generator
Automobile Tire Pump

Coal-fired Generators
Local Cogeneration
Small Hydro Generator
Nuclear Generators

Project 4.3:   Common Electrical Devices

Learning Expectations:

Pedagogical Issues

Can a student investigate an electrical device, and figure out how it works? This project requires exactly that.

Many of these are "hands on" devices, small items which can be taken apart and displayed.

This five day project could be modified to a library project, dealing with larger or less available items such as the nuclear generator or the fuel cell

Science Issues

In all cases, potential, current, and power are to be determined, or calculated.

The Learning Activity

Over a period of five days, students investigate a common electrical device to determine how it works.

Day 1   Choose a device to study. It should be small, available, and dissect - able.

Day 2   Draw a pictorial diagram, and follow the path of the current through the device. Draw a second, schematic diagram that deals only with the electrical components.

Day 3   Find any necessary quantities, and then calculate the potential, current, resistance and power of the device when it is operating properly.

Day 4   Write a step-by-step account of how the device works, starting at the point that the high-energy electrons enter the device, and ending at the point where the low-energy electrons leave the device.

Day 5   Write a full report, including Introduction, Diagrams, Calculations, How It Works, and Safety and Society issues, such as excess energy use or safety concerns.

Equipment, Preparation and Resources

Some worn, or discarded electrical devices. Students can often supply their own.

Categories:	Assessment and Evaluation
Knowledge:	
Inquiry:	quality of the investigation
Communication:	clarity and completeness of the writing
Applications, Extensions:	apprehension of energy use issues, consumer safety issues

Student Exercises

2 Explaining Electricity

Knowledge and Understanding

The first priority is to be able to recall the two theories emphasized in this unit. Every exercise requires you to use the theories to solve a problem. While most of the concepts are related to the theories, additional concepts will be introduced as needed.

You will also learn to make useful representations of electrical events . In particular, you will learn how do draw simple electrical circuit diagrams.

Knowledge and understanding are probed at regular intervals in the Grade Nine Daily quizzes. Study these as you go through the exercises, so that you can do your best when they are assigned.

Inquiry and Thinking

We will use the PEOE cycle for most labs and activities. You are expected to frame a question, provide your best prediction, and explain your thinking, using both sentences and diagrams.

At the end of the unit, you will be given a five day independent project. The project will demonstrate your ability to conduct your own investigation.

Communication

The quality of your arguments is the most important aspect of communication in this chapter. Your arguments consist of sentences, organized into paragraphs, and supported by diagrams or other representations.

Each sentence should be clear and to the point. You will find it best to limit your sentences to two concepts linked together to make a reasonable claim. If you need to relate more than two concepts, add a new sentence.

Applications, Connections and Extensions

Every exercise in this book is designed to support you as you learn appropriate theories and apply them to problems. In the labs, you demonstrate your understanding of a theory only by applying the theory. In the quizzes and projects, you are invited to make further connections and extensions of your learning.

Explaining Electricity

Introduction: Two Theories of Electricity

1. The **Electrical Model of Matter** derived from a brief study of static electricity.

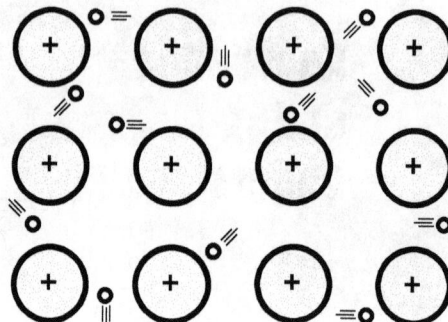

   1. **Atoms are made of a massive, positive nucleus surrounded by light, negative electrons.**
   2. **Positive and negative charges attract. Like charges repel.**
   3. **The outermost electrons in an atom can be made to move around.**
   4. **The nucleus and inner electrons are very difficult to move.**
   5. **Losing electrons makes matter positively charged.**
   6. **Gaining electrons makes matter negatively charged.**

   These ideas may already be familiar to you. They are the parts of the atom.

2. **A Theory of Energy and Electric Current** consists of six simple statements which can help you explain things that happen around you, and even help you to predict things you have never seen! Consider the following propositions of this theory.

   1. **Electrons can gain or lose electric potential energy.** An electron can be given potential energy by a dry cell, and lose that energy in a light bulb, for example.

   2. **All of the electrons in a single conductor have the same energy.** In a single copper wire, or copper wires connected directly together, all electrons inside will have the same energy.

   3. **Electrons move from high energy to low energy.** Like water or sand, electrons flow naturally from a position of high potential energy to a position of low energy.

   4. **Electrons must spend all of their energy in a circuit before returning to the energy source.** If a dry cell gives an electron 5 units of energy, the electron must spend all the energy before it returns to the dry cell to get more.

   5. **Resistance reduces the flow of electrons.** If you double the amount of resistance, the number of electrons that can get through in one second will be cut in half.

   6. **The more energetic the electrons, the greater number that get through a resistance.** If you double their energy, twice as many electrons will get through a resistance in one second.

   You will test some of the statements about electricity to see if they appear to be true. Then, you will use these statements to look at other experiments.

Introduction

These ideas can be formalized into a set of equations which can then be used to solve some electrical problems. You are unlikely to understand these equations unless you first understand the ideas above.

Single Component	Components in Series	Components in Parallel

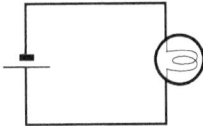

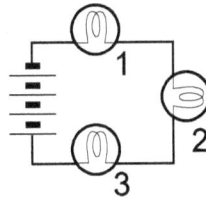

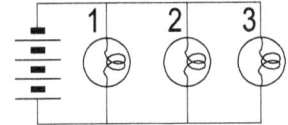

$$R = \frac{V}{I}$$

$$I = \frac{V}{R}$$

$$V = IR$$

$$P = VI$$

$$R_T = R_1 + R_2 + R_3 + \dots$$

$$I_T = I_1 = I_2 = I_3 = \dots$$

$$V_T = V_1 + V_2 + V_3 + \dots$$

$$P_T = P_1 + P_2 + P_3 + \dots$$

$$\frac{1}{R_T} = \frac{1}{R_1} + \frac{1}{R_2} + \frac{1}{R_3} + \dots$$

$$I_T = I_1 + I_2 + I_3 + \dots$$

$$V_T = V_1 = V_2 = V_3 = \dots$$

$$P_T = P_1 + P_2 + P_3 + \dots$$

The equations look difficult, but a little experience with them will help you understand how they relate to real circuits.

In all of the exercises in this book, the question must be answered in *complete sentences*. One sentence is one thought. A single word is simply not enough.

Explaining Electricity

Lab 1.1: Charging Up

What's The Question?

How do you charge a plastic strip? Can you transfer a charge to another object?
Do charged objects exert forces upon each other? How do we predict these forces?

What Are We Doing?

You will be using *vinylite* and *acetate* strips, paper, fur, and two *pith balls*: A and B.

Before each experiment, **predict** the outcome, and **explain** your reasons for your prediction. During the experiment, **observe** and draw a picture of your observations. Finally, **explain** your observations, making changes in your explanation from what you learned.

1. Rub an acetate strip with paper. Bring the charged acetate strip near pith ball A, *allow it to touch*, and observe what happens.

2. Rub a vinylite strip with fur. Bring the charged vinylite strip near a pith ball B, *allow it to touch*, and observe what happens.

3,4. Bring each charged strip near the opposite pith ball. *Do not allow to touch.*

(If they do touch, discharge everything by rubbing it with foil, and then start over again at 1.)

5. Bring pith balls A and B close to each other. *Do not allow them to touch.*

What Are We Thinking About?

1. To charge an object, you must either add or remove electrons from it.

2. *Ebonite* or *Vinylite* will gain electrons from *fur*.

3. *Acetate* will lose electrons to *paper*.

4. Matter is made of atoms. Each atom contains a massive, positive *nucleus* surrounded by light, negative *electrons*.

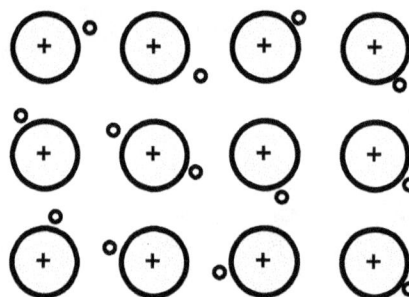

5. The outermost electrons can sometimes be made to move around within the solid. The nucleus and its inner electrons are not able to move around.

Questions For Later...

1. Do like charges *attract* each other, or *repel* each other? Explain your evidence.

2. Do opposite charges *attract* each other, or *repel* each other? Explain your evidence.

3. How many different kinds of electrical charge have you seen? Are those the only kinds?

Matter and Static Electricity

Name:

Date:

1 *Predict, Explain:* Acetate near pith ball A

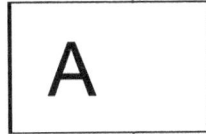

Observe and Explain

(A) [A]

2 *Predict, Explain:* Vinylite near pith ball B

Observe and Explain

(B) [V]

3 *Predict, Explain:* Acetate near pith ball B

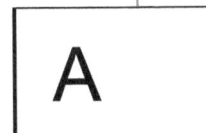

Observe and Explain

(B) [A]

4 *Predict, Explain:* Vinylite near pith ball A

Observe and Explain

(A) [V]

5 *Predict, Explain:* Pith ball A and B

Observe and Explain

(A) (B)

Explaining Electricity

Lab 1.2: Electrons and Charge

What's The Question?

Metals are shiny, good conductors of heat, good conductors of electricity, and can usually be bent. Non-metals are dull in appearance, poor conductors of heat, poor conductors of electricity, and cannot usually be bent. Electrons behave very differently in the two kinds of matter.

How can we use our knowledge of matter to create and transfer static electric charges?

What Are We Doing?

1. Complete the exercises on the back of this sheet. Count all of the electrons and positive charges on each body, and add the charges to find the net positive or negative charge on each body.

2. You will be given acetate and vinylite strips, and metal covered pith balls. Suspend the pith balls so that they swing freely. Touch them with your finger to make them neutral.

3. Charge the acetate positively by rubbing it with paper; charge the vinylite strip negatively by rubbing it with wool.

4. Bring the charged strips close to the pith balls, and watch what happens when they touch. Draw particle diagrams to show what is happening with the electrons.

5. Your teacher will use a high voltage Van de Graaf machine to demonstrate how to charge a metal bar.

What Are We Thinking About?

1. Non-metal atoms are small, with many positive charges. Their electrons are trapped in tightly held electron pairs between the atoms.

2. Metal atoms are large, with few positive charges. Their weakly held electrons travel quite freely within the metal crystal.

3. Mobile electrons anywhere are able to reflect light; trapped electrons cannot.

4. Mobile electrons can conduct heat energy; trapped electrons cannot.

5. Mobile electrons can conduct electric charge; trapped electrons cannot.

6. When metals are bent, the electrons and atoms just slide into new positions. When non-metals are bent, the lattice arrangement breaks into pieces.

Questions For Later...

1. You bring a positively charged strip up to a pith ball. The pith ball moves away from the positive strip. What is the charge on the pith ball? Explain your answer.

2. As you are driving your car, it picks up an electrical charge by rubbing on the road and the air. You don't like to get a shock when you get out of your car, so you attach a small metallic strip to the frame of the car, so that it drags on the ground. How would it prevent shock?

Matter and Static Electricity

What charge is on the pith ball and the vinylite strip? Would these attract or repel each other?	Redraw the pith ball and the strip after they have touched. Do they attract or repel each other? Explain.
What charge is on the pith ball and the acetate strip? Would these attract or repel each other?	Redraw the pith ball and the strip after they have touched. Do they attract or repel each other? Explain.
What charge is on the pith ball and the vinylite strip? Would these attract or repel each other?	Redraw the pith ball and the strip after they have touched. Do they attract or repel each other? Explain.
What charge is on the pith ball and the vinylite strip? Would these attract or repel each other?	Redraw the pith ball and the strip after they have touched. Do they attract or repel each other? Explain.
What charge is on the pith ball? Would the electrons in the conducting bar be attracted or repelled? Draw the electrons.	The conducting bar has been allowed to make brief contact with the ground. Redraw the electrons in the conducting bar.
What charge is on the pith ball and the vinylite strip? Would these attract or repel each other?	The conducting bar has been allowed to make brief contact with the ground. Redraw the electrons in the conducting bar.

Quiz 1.3: Matter and Static Electricity

1 These diagrams refer to two different kinds of matter, **A** and **B**. Choose **A**, **B**, or Neither to answer the following questions.

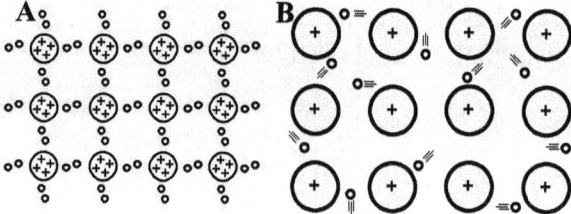

_____ easily bent and hammered into new shapes
_____ likely to be shiny
_____ likely to be a poor conductor of electricity
_____ likely to be a good conductor of heat
_____ best described as "non-metal"

Date: _____ / 4

2 Add electrons to each diagram as needed to provide an overall charge of +4 on A and -4 on B.

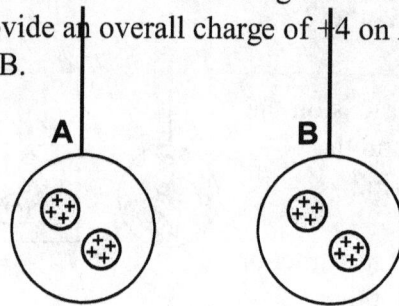

Do these pith balls *attract* or *repel* each other? Give reasons for your answer.

Date: _____ / 4

3 What is the overall charge on each pith ball A and B?

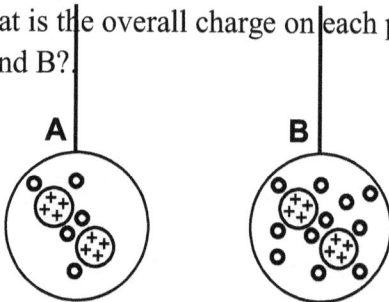

Do these pith balls *attract* or *repel* each other? Give reasons for your answer.

Date: _____ / 4

4 What is the overall charge on each pith ball A and B?

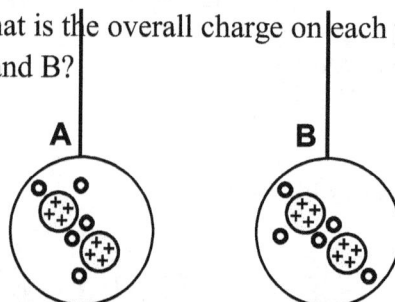

Do these pith balls *attract* or *repel* each other? Give reasons for your answer.

Date: _____ / 4

Quiz 1.3: Matter and Static Electricity   **Name:**

5 On pith ball A, draw an excess of 5 electrons. On pith ball B, draw a deficiency of 3 electrons.

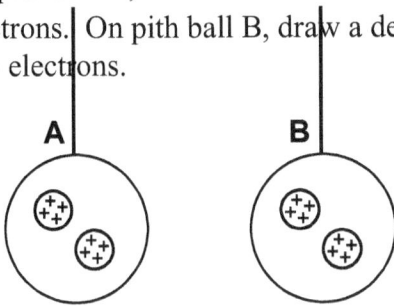

A   **B**

_____ What is the charge on pith ball A?
_____ What is the charge on pith ball B?
Do these pith balls attract or repel each other?
Give reasons for your answer.

Date: / 4

6 Both the pith ball and the charged strip are coated with a thin metal conductor. They are brought into contact.

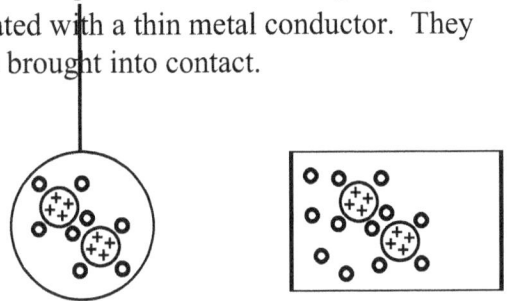

Change the diagram to show any movement of electrons.
After they contact, do they *attract* or *repel* each other? Give reasons for your answer.

Date: / 4

7 On pith ball A, draw a deficiency of 4 electrons. On pith ball B, draw a deficiency of 3 electrons.

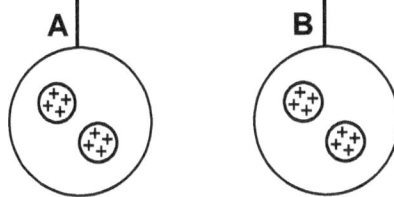

A   **B**

_____ What is the charge on pith ball A?
_____ What is the charge on pith ball B?
Do these pith balls attract or repel each other?
Give reasons for your answer.

Date: / 4

8 A charged ball is brought close to, but not touching, a neutral metal rod. Draw the new positions of the electrons in the metal rod.

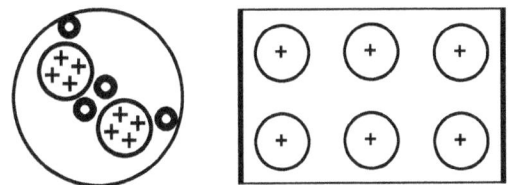

_____ What is the charge on the pith ball?
_____ What is the charge on the metal rod?

What would happen if these objects touched?

Date: / 4

Lab 2.1: Getting a Light Bulb to Glow

What's The Question?

You will be given a light bulb, a flashlight dry cell, aluminum foil, copper wire, paper, tape and some other stuff. (Perhaps you won't need to use it all). *How will you arrange these things in order to make the light bulb light up?*

What Are We Doing?

1. Predict: draw a diagram of your plan.

2. Explain how your plan is supposed to work. Write complete sentences and draw a diagram.

3. Follow your plan to see if it works. If it does not work keep trying different ways until it does work. Observe what works and make another picture of the working arrangement.

4. Explain how it works using your own words and complete sentences.

What Are We Thinking About?

1. What is a dry cell (battery) and how does it work?

2. Which of the materials you were given are metals and which are non-metals?

3. Does a light bulb have parts? What do you think they might be like?

Questions For Later...

1. Check three other peoples' successful arrangements. What appears to be the same in each of them?

2. Check other peoples' unsuccessful arrangements. What appears to be the problem?

3. When electrical particles (electrons) finish going through the bulb, must they go back to the dry cell? Explain your thinking.

Energy and Electric Current

Name:

Date:

K I C A

Focus Question: Write the question that you are trying to answer.

---

1 **Predict:** What arrangement of the materials would allow the light bulb to light up? Draw any materials that you think that you would need.

2 **Explain** your thinking. Use complete sentences to show how your plan should work.

---

3 **Observe,** and record your observations here. If your plan does not work, then keep working until it does, and draw a new picture.

4 **Explain** how your experiment works using complete sentences.

Lab 2.2: What's Inside a Light Bulb?

What's The Question? What parts are inside a light bulb? Everyone has had a collection of things, like coins or CD's. People sort their collections into groups or *categories*. Coins can be separated by country, by date, by metal, or by other properties.

If you take a light bulb apart, how would you group the parts into two or three categories of things?

What Are We Doing?

1. *Predict*: From whatever memories you have, draw a picture of the insides of a light bulb. Label the parts, and *explain* what they do.

2. ***Your teacher may do this step!*** Put a light bulb inside a plastic bag. Put that bag inside another bag. Place the bulb on a hard surface. Whack the bulb with a small hammer or a file, just hard enough to break the glass.

3. Carefully shake all the parts out of the bulb out into a tray or dish.

4. Using forceps, separate the parts into categories that you think are appropriate. Make up a name for each part, and category.

5. Make a record of your *Observations*. A labeled picture would be good. *Explain* how a light bulb works.

What Are We Thinking About?

1. Broken glass is a hazard. Wear your safety glasses at all times. Do not handle broken glass with your hands.

2. Through which part or parts of the bulb does electricity actually pass?

3. Which part or parts of the bulb keep electricity away from other parts?

4. How does the electricity get from the energy source (dry cell or household current) outside the bulb to the parts inside the bulb that actually light up?

Questions For Later...

1. What is a *conductor*? What appearances do conductors have in common?

2. What is an *insulator*? What appearances do insulators have in common?

3. When a light bulb burns out, what part or parts are affected? Exactly what happens to them, and how does that keep the bulb from working?

Energy and Electric Current

Name:

Date:

K I
C A

Focus Question: Write the question that you are trying to answer.

---

1 **Without looking, Predict** what you believe the inside of a light bulb looks like

2 **Explain** how you think your light bulb works.

3 **Observe,** and record your observations here.

4 **Explain** how you believe the light bulb works.

Explaining Electricity

Lab 2.3: What Makes Light Bulbs Different?

What's The Question?

When a different bulb is used, you get different results. Some bulbs are brighter, others are dimmer.

What is it about different light bulbs that makes them bright or dim?

What Are We Doing?

You will be given one cell, some wires, and two different bulbs.

1. Examine the bulbs carefully. Predict which bulb you think will be the brightest, and use a picture to show how you would arrange the wires to make the bulb light up.

2. Explain your prediction, using complete sentences and more pictures.

3. Observe the results of your experiment and make a labeled picture and use complete sentences describing what you observed.

4. Explain your results in complete sentences. Use more pictures if necessary.

What Are We Thinking About?

There are two lakes, connected by a short stream. Lake Alan is 10 m higher than Lake Betty. Between the two lakes is a dam that holds the water back, and keeps Lake Alan from flowing into Lake Betty. Every once in a while, workers open a gate at the bottom of the dam so that some water can rush out. The gate can be opened just a little or quite a lot.

1. In which case does the gate offer the most resistance to the water: just a little open, or quite a lot open?

2. In which case is more energy (sound and motion) released: when the gate is just a little open, or quite a lot open?

Questions For Later...

1. You can put either a 60 W bulb or a 100 W bulb into the same lamp at home. Does that change the energy of the electrons coming out of the wall socket?

2. Does each electron that goes through a 60 W bulb have more energy, less energy, or the same energy to spend as the electron going through a 100 W bulb?

3. Compare the resistance in the two bulbs. Which bulb has the greatest resistance: the 60 W bulb, or the 100 W bulb?

Energy and Electric Current

Focus Question: Write the question that you are trying to answer.

1 **Predict:** Which bulb will produce the most light?

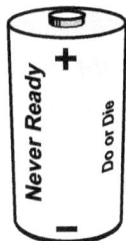

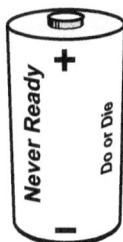

2 **Explain:** why you thought one bulb would be brighter than the other.

3 **Observe,** and record your observations here.

4 **Explain** your observations using what you have learned.

Lab 2.4: Getting Two Identical Bulbs to Glow

What's The Question?
You will be given two identical light bulbs, one dry cell, and some wires.

How many distinctly different ways can you arrange them so that both the light bulbs light up at the same time? Will there be any difference in the brightness of the bulbs in each case?

There is more than one way to do this!!

What Are We Doing?

1. **Predict** a number of working arrangements by drawing pictures of your plans.

2. **Explain** how your plans work, using complete sentences.

3. Construct the arrangements you planned, to see if they work. If they do not work, keep trying different ways until you do get them to work.

4. **Observe** your working arrangements, and make pictorial records of them.

5. **Explain** how your arrangements work. In particular, try to explain any differences in the brightness of the light bulbs.

What Are We Thinking About?

1. In each plan, how will the electricity get to the light bulb?

2. What happens to the light bulbs if one of the wires becomes unconnected? That is called a *break in the circuit*, or an *open circuit*.

3. How could you begin to draw pictures of the electricity in your arrangements, so that you could begin to explain what is happening?

Questions For Later...

1. Can electricity go forward and backward along the same wire at the same time? Draw pictures and write complete sentences explaining your thinking.

2. If electricity reaches a place where two wires are joined, what does it do? Explain.

3. In which of your working circuits would the battery go dead first? Explain.

Energy and Electric Current

Name:

Date:

Focus Question: Write the question that you are trying to answer.

1   **Predict** how you can use wires to make both light bulbs light up in different ways.

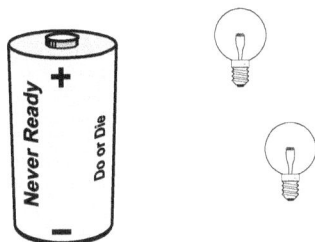

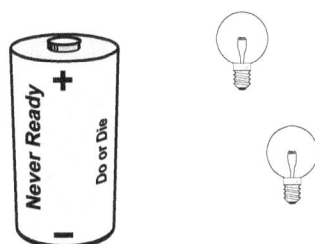

2   **Explain** how your plan will work, especially how the electricity will get to the light bulbs.

3   Experiment and **Observe** how different arrangements work. Make records (pictures) of your working models.

4   **Explain** your observations. Are the bulbs the same brightness in all arrangements? Why or why not?

Explaining Electricity

Lab 2.5: Getting Two Different Light Bulbs to Glow

What's The Question?

You can get two identical light bulbs to light up in two distinctly different ways.

In how many different ways can you get two different light bulbs to light up?
Are they brighter, dimmer, or the same as they were in Lab 1.2.4 ?

What Are We Doing?

1. **Predict** at least two different working arrangements containing two different bulbs. Draw pictures.

2. **Explain** how your arrangements are supposed to work, especially how the electricity gets to the light bulbs.

Experiment with your arrangements. If they don't work, keep trying until you do get them to work.

3. **Observe** your working arrangements, and make records (pictures and words).

4. **Explain** how the working arrangements actually work.

What Are We Thinking About?

1. Will the lamp closest to the bulb always be brighter than it would be if it was farther away?

2. Does the "downstream" bulb have any effect upon the "upstream" bulb? Explain

3. What is getting "used up" in these arrangements? The light bulb? Electrons? Energy? Matter in the dry cell?

Questions For Later...

1. After the *electrons* have moved from the dry cell through the bulbs making them glow, must they all go back to the dry cell? Explain your answer.

2. After the *energy* has moved from the dry cell, to the light bulbs, does it go back to the dry cell? Explain your answer.

3. On a fresh sheet of paper, draw at least four different circuits from the previous Lab 1.6 and this lab. Compare the energy released by the bulbs in each case. Explain why you believe the energies released are what they are.

Energy and Electric Current

Name:
Date:

K I C A

Focus Question: Write the question that you are trying to answer.

1 **Predict** how many different ways you can get two different light bulbs to glow. Then predict how bright the bulbs will be in each case.

2 **Explain** how your plans will work, especially why you believe the bulbs will be bright or dim.

3 **Observe,** and record your observations here.

4 **Explain** your observations. Is each light bulb the same brightness as it was in Lab 1.2.4?

Quiz 2.6: Energy and Electric Current

1 Name the objects represented in this picture:

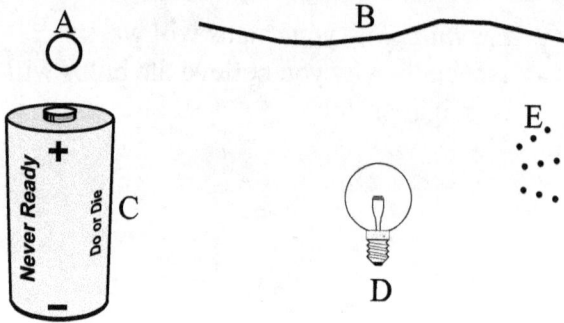

A _____ D _____

B _____ E _____

C _____

Date: _____ / 4

2 Name the objects represented in this picture:

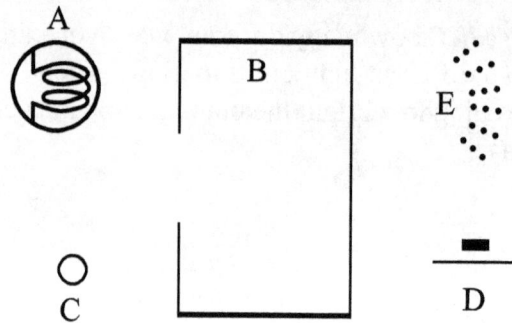

A _____ D _____

B _____ E _____

C _____

Date: _____ / 4

3 Complete the diagram to determine how many electrons pass through the bulb, and how much energy they lose there.

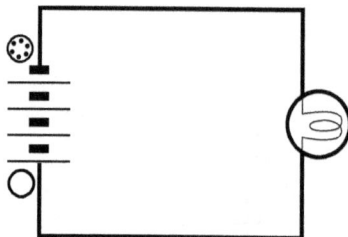

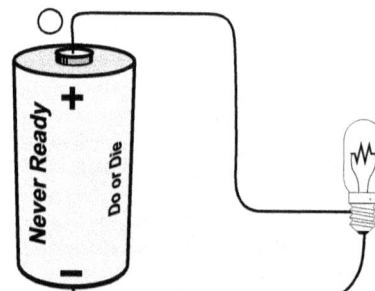

Resistance = 3 units

Date: _____ / 4

4 Complete the diagram to determine how many electrons pass through the bulb, and how much energy they lose there.

Resistance = 6 units

Date: _____ / 4

5   Complete the diagram to determine how many electrons pass through the bulb, and how much energy they lose there.

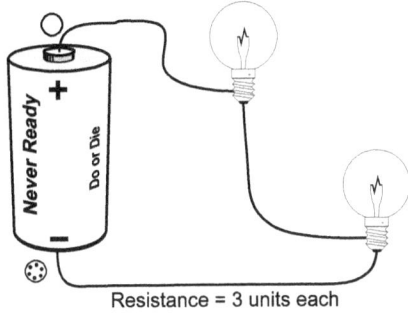

Resistance = 3 units each

Date: / 4

6   Complete the diagram to determine how many electrons pass through the bulb, and how much energy they lose there.

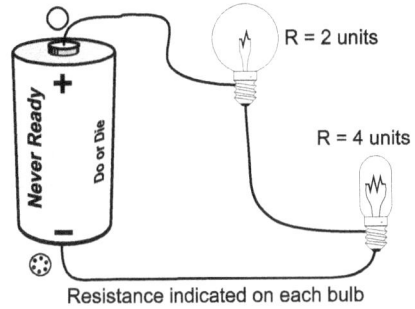

R = 2 units

R = 4 units

Resistance indicated on each bulb

Date: / 4

7   Complete the diagram to determine how many electrons pass through the bulb, and how much energy they lose there.

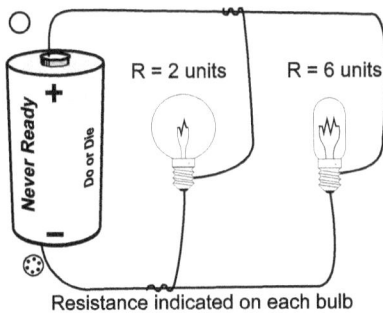

R = 2 units   R = 6 units

Resistance indicated on each bulb

Date: / 4

8   Complete the diagram to determine how many electrons pass through the bulb, and how much energy they lose there.

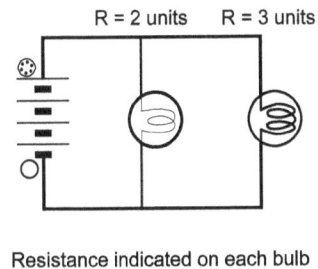

R = 2 units   R = 3 units

Resistance indicated on each bulb

Date: / 4

Activity 3.1: Scientists' Ideas of V, I, R, & P

Do you Remember? List the 6 propositions of the Theory of Energy and Electric Current

1. _____
2. _____
3. _____
4. _____
5. _____
6. _____

What's The Question?
We have been exploring current electricity for several days. Until now, we have talked about electricity in everyday terms. How do scientists talk about electricity?

What is the meaning of the words potential, current, resistance, *and* power?

What Are We Doing?
1. Write your own ideas about each group of words on the opposite page.(5 min)

2. Get into home groups of four students. Share your ideas. Circle the ideas about which you disagree. (10 min)

3. Assign each person from your group to one of the expert groups. Take your paper with you to learn what you can about your topic. (30 min)

4. Return to your home group and share your ideas. (30 min)

5. Summarize your learning on the opposite page. Use additional paper if necessary.

What Are We Thinking About?
1. How does water behave when it flows from higher to lower levels in a ditch or a hose?

2. How do people behave when they are spending a limited amount of money in a shopping mall?

3. What could cause resistance in a stream of water?

4. What could cause resistance in a line of shoppers at a checkout?

Questions For Later...
1. A toaster has a heating wire with 20 Ω resistance connected across a potential of 120 V. How much current will pass through the wire?

2. What is the power of that wire, in Watts?

Energy and Electric Current

Name:

Date:

Focus Question: Write the question that you are trying to answer.

Volts, voltage, electric potential	*Amperes, current, amperage*
Ohms, resistance, resistor	*Watts, power, kilowatts*

Potential is the ability to do work.

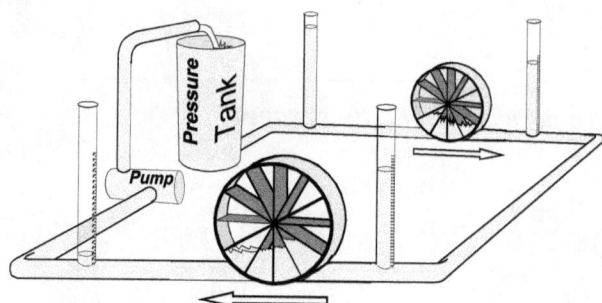

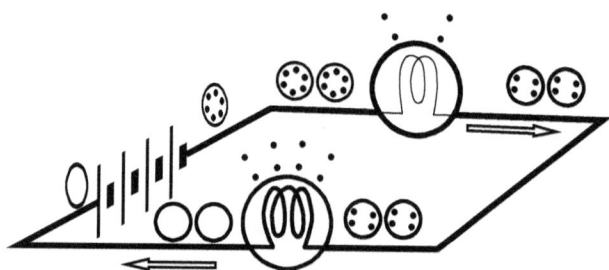

Have you ever held your thumb over the end of a hose? When you press hard with your thumb, you create a lot of resistance. Do you feel how the pressure of the water increases? The water has to spend more of its potential to get past your thumb.

The equation **V = IR** means that more potential energy is spent when a lot of electric current passes through a lot of resistance.

Electric Potential

Ideas

Electric potential is like water level: the higher the level, the greater the potential. A pump pushes on the water, lifting it to a high level. The water in the pipe is then under pressure. You can see the water level in the four vertical tubes.   At the first water wheel, the water converts some of its potential into the mechanical energy of the turning wheel.   The water level (and the pressure) are decreased. At the second water wheel, the water spends its remaining potential until the pressure reaches zero. The water must return to the pump before it can do more work.

Electric potential is like money: the more money in your pocket, the greater your potential. When you reach the first checkout, you spend some of your potential to buy a ticket. At the second checkout, you spend your money until your pockets are empty, then you go back to the bank for more.

Electric potential is *energy per electron*. A dry cell gives each electron (the circles) a small amount of energy (the dots).   At the first light bulb, the electrons convert some of their energy into other forms. At the second light bulb, the electrons spend their energy until they reach a potential of zero. They can do no more work until they return to the dry cell for more energy.

Metric System of Units: One *Volt* equals one *Joule per Coulomb*. A coulomb is a very large number of electrons, about two million million million. Six volts means that every single coulomb has six joules of energy.

Mathematical Equations

$$V = IR$$

Energy and Electric Current

Name:

Date:

Electric Current is the flow of electrons.

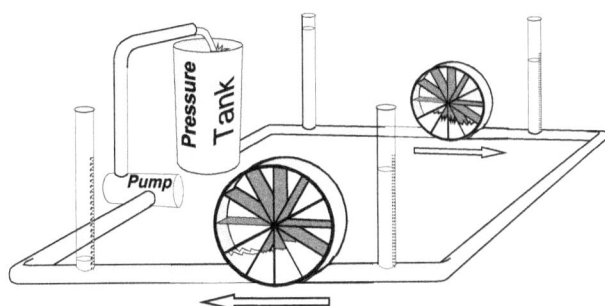

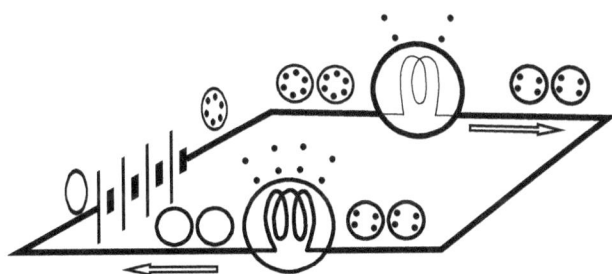

Have you ever dug a ditch in a sandbox to drain a puddle? The ditch must flow from high to low. The greater the potential, the greater the flow of water. If you plop some sand in the ditch, the water slows down and backs up. Perhaps it backs up enough to go around the resistance!

The equation **I = V/R** means that increasing the potential difference **V** causes more current **I** to flow. More resistance **R** causes less current to flow.

Electric Current

Ideas

Electric current is like flowing water: when something supplies a difference in height, the water must flow from higher potential to lower potential. All of the water flowing into a pipe or water wheel must come out. Water wheels slow the flow of water, and take energy. The resistance of the water wheel slows the flow of the water, and causes the water to "pile up" behind the resistance.

Electric current is like a crowd of people moving through a shopping mall. No one moves unless they are given some money. The more money they get, the quicker they get going! People flow from rich to poor. Everyone who goes into an aisle or a shop comes out again. Checkout counters slow the people, and take their money. The checkout resistance causes people to "pile up" into lines. If the line gets very long, people are willing to spend more money to get through.

Electric current is a flow of electrons in a circuit. If there is a potential difference, the electrons will begin to move from high to low potential through the conductors. All electrons that enter a resistor or conductor must come out; electrons cannot be created or destroyed. Light bulbs slow the flow of electrons, and convert their energy to heat and light. The resistance of the light bulbs causes electrons to "pile up," increasing the potential at that point. The more potential they have, the faster the flow of electrons through the resistance.

Metric System of Units One *Ampere* is a flow of *one coulomb per second*, or about two million million million electrons per second.

Mathematics

$$I = \frac{V}{R}$$

Explaining Electricity

Resistance costs flowing electrons energy.

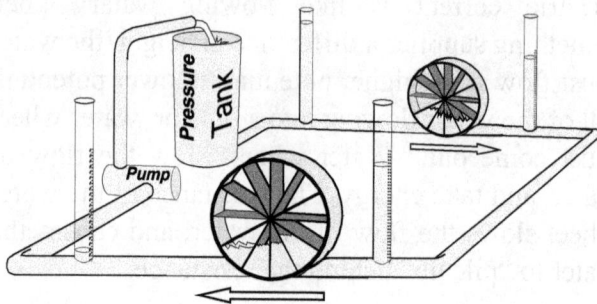

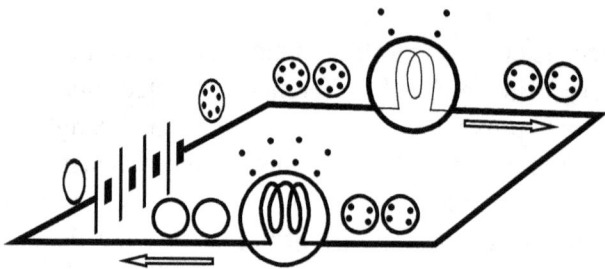

Two people let a rope slide through their hands as they lower a bucket. That's like two resistances to the motion of the rope. If one person reduces his resistance, then more rope, and more energy, is spent on the other person's hand.

The equation **R = V/I** means that if we want to know how big a resistance is, we put a potential difference **V** across it, and then observe how much current **I** flows. Small **I** means great resistance.

Electric Resistance

Ideas

Resistance is like a water wheel. Water must spend some energy to get through the wheel. The resistance turns the potential energy of height into other forms of energy. When the water comes out of the resistance, some of its potential is gone. When two water wheels are placed as shown, the same water must flow through both. Whatever potential it does not spend in the first resistance, it must spend in the second.

Resistance is like checkout counters: they slow you down, and take your money. These checkout counters are odd. They behave more like ticket outlets: everyone gets a chance to get a ticket, but only a few tickets are sold each minute. If you are allowed to buy a ticket, you still have to pay. You can imagine how much more you would be willing to pay if your chances of getting a ticket were slim, and how much less you would pay if tickets were easy to get. In this picture, the shoppers are trying to buy two different tickets.

Resistance causes electrons to bumble, bounce and stick to atoms as they go through. The random bumbling and bouncing causes the atoms in the resistance to move more quickly, that is, to heat up. Resistance converts electrical potential into heat energy! The greater the resistance, the fewer electrons go through. The amount of energy actually converted depends upon the *whole* circuit.

Metric System of Units One *Ohm* is the resistance that will allow one *Ampere* of current to flow when one *Volt* of potential is applied.

Mathematics

$$R = \frac{V}{I}$$

Energy and Electric Current

Power is the rate at which work is done.

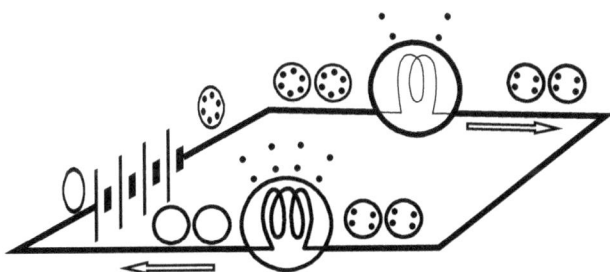

Electric Power

Ideas

The rate at which work is done in each water wheel depends upon two things. The greater the potential difference, the more work is done by every drop of water. The greater the flow of water, the faster each drop does its work on the water wheel.

Power is like the total amount of money spent on tickets in a given day. The rate at which money is spent in each checkout counter depends upon two things. The greater the price of each ticket, the more money is spent. At the same time, the more people who buy tickets, the faster the money is spent.

Power is the rate at which electrons spend their energy in a resistance. Power depends upon two things. The greater the potential difference across a resistance, the greater the energy converted to heat and light. The more electrons passing through a resistance, the faster the energy is converted.

Metric System of Units The *Watt* is the rate of converting energy from one form to another. One Watt is a rate of conversion of *one Joule per second*.

Mathematics

$$P = VI$$

$$P = I^2R$$

$$P = \frac{V^2}{R}$$

The equation *P = VI* means that power depends upon both the voltage and the current. When you turn up the dimmer switch on a 100 W light bulb, you increase the potential (volts). That increases the current (amps). More electrons go through the bulb, and each electron spends more energy. The bulb gets much hotter and brighter.

Explaining Electricity

Lab 3.2: Series and Parallel Circuits

Do you Remember? List the 6 propositions of the Theory of Energy and Electric Current

1. _____
2. _____
3. _____
4. _____
5. _____
6. _____

What's The Question? You have been experimenting with different ways to connect identical light bulbs together so that they both light up (Lab 1.2.4). You have also been playing with arranging two different bulbs in different ways (Lab 1.2.5). *What makes these different circuits behave the way that they do?*

What Are We Doing?

1. Obtain four stir sticks, some chewing gum and a balloon.

2. Inflate the balloon. Place two books on the desk so the balloon just touches both books.

3. Secure one stir stick in the mouth of the balloon, and time it as the balloon deflates.

4. Combine the stir sticks with chewing gum in various ways, (see opposite).

5. Inflate the balloon to the same degree, and time the balloon as it deflates through each combination.

   "Is the air flow greater than, less than, or equal to the air flow in the single stir stick? Which combination has the greatest resistance?"

What Are We Thinking About?

Two devices are in series if everything that flows through the first must then flow through the second (the flights of stairs are in series).

Two devices are in parallel if they provide two different paths from one starting point to a single ending point. (The fire pole is in parallel with the stairs)

Questions For Later...

1. Are hot and cold water taps in your kitchen *in series* or *in parallel*?

2. Are the hills in a road *in series* or *in parallel*?

3. Which combination of pairs of straws made the greatest resistance: *series* or *parallel*?

4. Which combination of pairs of straws made the least resistance: *series* or *parallel*?

Series and Parallel Circuits

Focus Question:

Before you start, **predict** the air flow, and **explain** your prediction.
After each trial, record your **observations**, and **explain** any differences from your predictions.

1 *Two stir sticks in series:*

2 *Four stir sticks in series:*

3 *Two stir sticks in parallel*

4 *Four stir sticks in parallel*

Activity 3.3: Calculating V, I, R and P

> *Do you Remember?* Write the Equations for V, I, R and P

Single Components Every component in a circuit has a potential causing a current to flow through a resistance, releasing energy or power. For each component, V, I, R and P can be calculated using the equations you wrote above. Try it!

$V_T =$
$I_T =$
$R_T =$
$P_T =$

100 V

$V_1 =$
$I_1 =$
$R_1 = 250 \; \Omega$
$P_1 =$

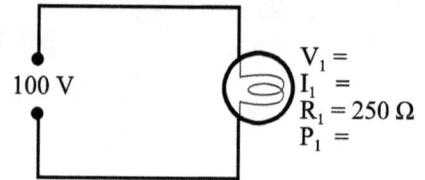

Series Components

$$R_T = R_1 + R_2 + R_3 + \ldots$$

$$I_T = I_1 = I_2 = I_3 = \ldots$$

$$V_T = V_1 + V_2 + V_3 + \ldots$$

$$P_T = P_1 + P_2 + P_3 + \ldots$$

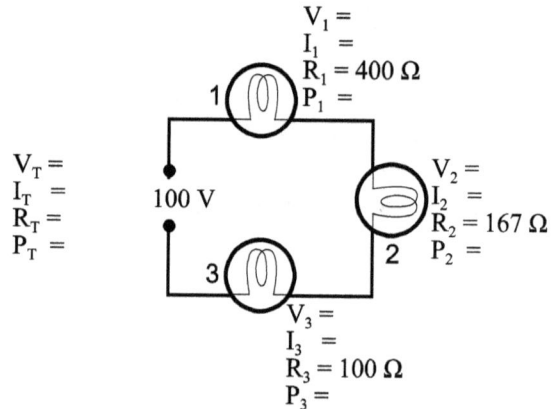

$V_T =$
$I_T =$
$R_T =$
$P_T =$

100 V

1

$V_1 =$
$I_1 =$
$R_1 = 400 \; \Omega$
$P_1 =$

2

$V_2 =$
$I_2 =$
$R_2 = 167 \; \Omega$
$P_2 =$

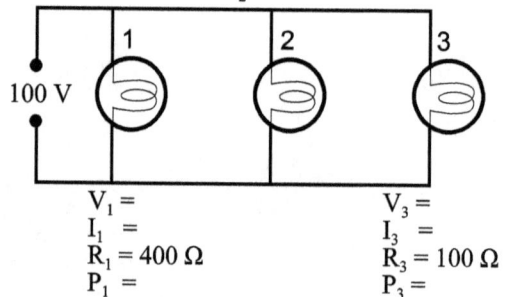

3

$V_3 =$
$I_3 =$
$R_3 = 100 \; \Omega$
$P_3 =$

Parallel Components

$$\frac{1}{R_T} = \frac{1}{R_1} + \frac{1}{R_2} + \frac{1}{R_3} + \ldots$$

$$I_T = I_1 + I_2 + I_3 + \ldots$$

$$V_T = V_1 = V_2 = V_3 = \ldots$$

$$P_T = P_1 + P_2 + P_3 + \ldots$$

$V_T =$
$I_T =$
$R_T =$
$P_T =$

$V_2 =$
$I_2 =$
$R_2 = 167 \; \Omega$
$P_2 =$

100 V

1   2   3

$V_1 =$
$I_1 =$
$R_1 = 400 \; \Omega$
$P_1 =$

$V_3 =$
$I_3 =$
$R_3 = 100 \; \Omega$
$P_3 =$

Series and Parallel Circuits

Use the equations on the previous page to solve for V, I, R, and P for each component in the circuit, and totals for the whole circuit.

1

$V_T =$
$I_T =$
$R_T =$
$P_T =$

100 V

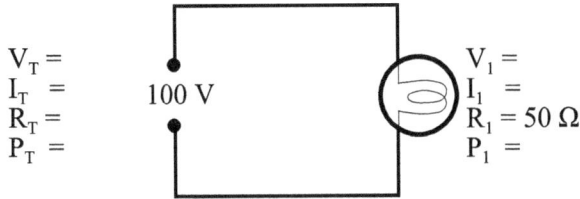

$V_1 =$
$I_1 =$
$R_1 = 50\ \Omega$
$P_1 =$

2

100 V

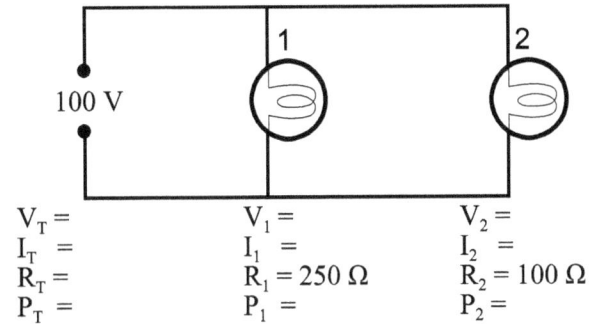

$V_T =$
$I_T =$
$R_T =$
$P_T =$

$V_1 =$
$I_1 =$
$R_1 = 250\ \Omega$
$P_1 =$

$V_2 =$
$I_2 =$
$R_2 = 100\ \Omega$
$P_2 =$

3

$V_T =$
$I_T =$
$R_T =$
$P_T =$

100 V

$V_1 =$
$I_1 =$
$R_1 =$
$P_1 = 500\ W$

4

$V_T =$
$I_T =$
$R_T =$
$P_T =$

100 V

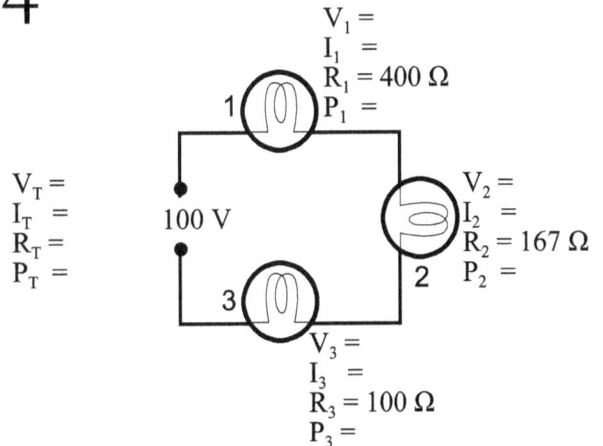

$V_1 =$
$I_1 =$
$R_1 = 400\ \Omega$
$P_1 =$

$V_2 =$
$I_2 =$
$R_2 = 167\ \Omega$
$P_2 =$

$V_3 =$
$I_3 =$
$R_3 = 100\ \Omega$
$P_3 =$

5

100 V

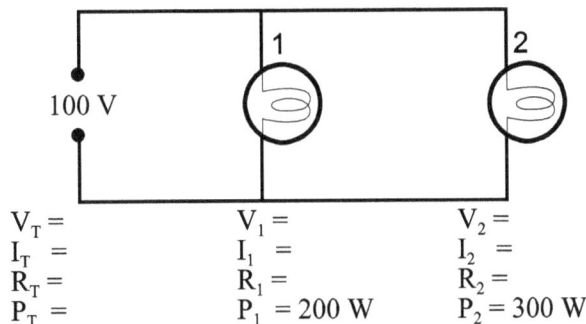

$V_T =$
$I_T =$
$R_T =$
$P_T =$

$V_1 =$
$I_1 =$
$R_1 =$
$P_1 = 200\ W$

$V_2 =$
$I_2 =$
$R_2 =$
$P_2 = 300\ W$

6

$V_T =$
$I_T =$
$R_T =$
$P_T =$

100 V

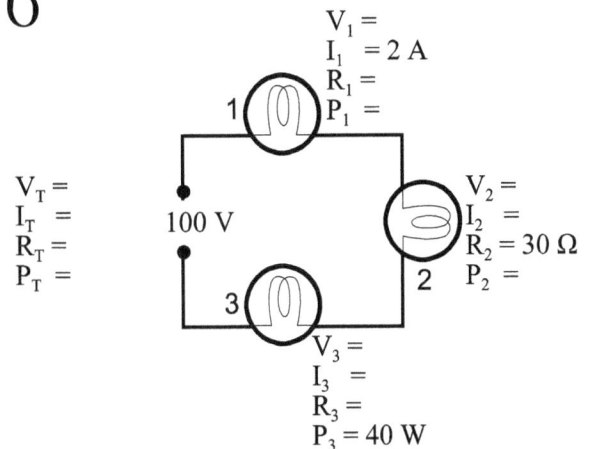

$V_1 =$
$I_1 = 2\ A$
$R_1 =$
$P_1 =$

$V_2 =$
$I_2 =$
$R_2 = 30\ \Omega$
$P_2 =$

$V_3 =$
$I_3 =$
$R_3 =$
$P_3 = 40\ W$

Quiz 3.4: Simple Electric Circuits

1 A water circuit has two pinch clamps at P and Q , which cause the water levels to stay as indicated in pipes A, B, C, and D.

If Q is tightened, what will happen to the water levels? (Rise, fall, stay the same)

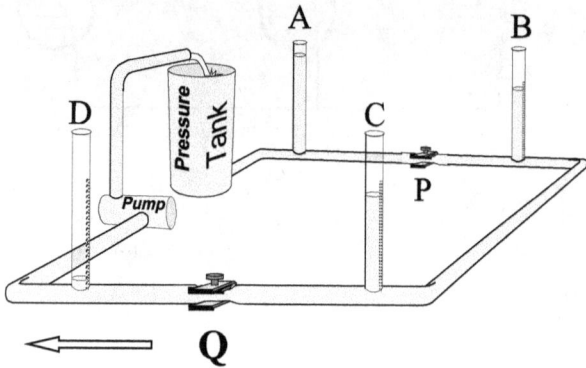

Draw the new water levels on A, B, C, and D

Date: _____ / 4

2 A water circuit has two pinch clamps at P and Q , which cause the water levels to stay as indicated in pipes A, B, C, and D.

If P is loosened, what will happen to the water levels? (Rise, fall, stay the same)

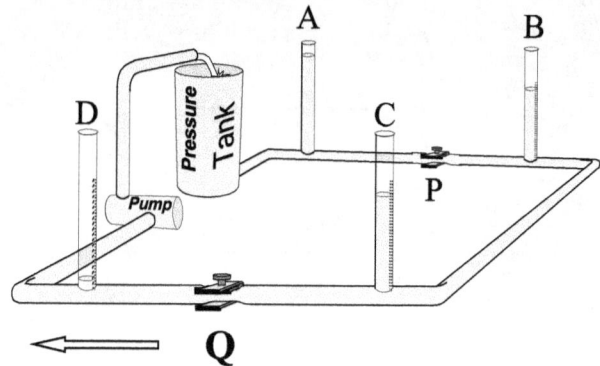

Draw the new water levels on A, B, C, and D

Date: _____ / 4

3 Each person gets $100 as they pass through the bank, and they must spend it all.

If each person spends $25 at Q, How much money do they have at A, B, C, and D?

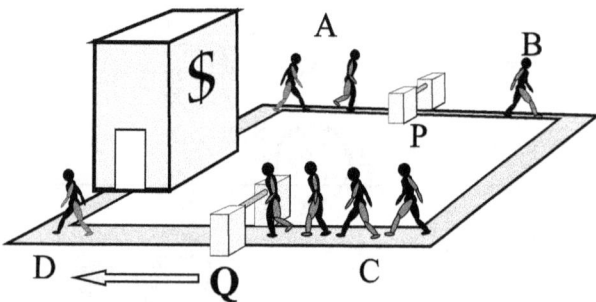

Write the amount of money at A, B, C, and D

Date: _____ / 4

4 Everyone who passes through the bank gets $50.

If each person spends $10 at P, how much money do they have at A, B, C, and D?

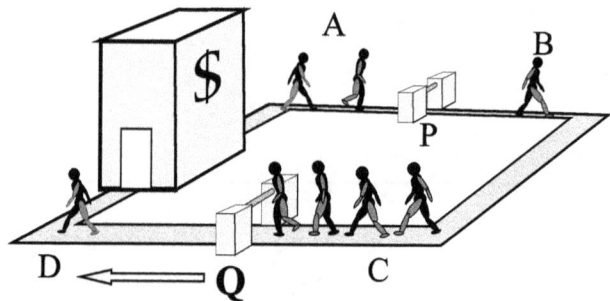

Write the amount of money at A, B, C, and D

Date: _____ / 4

5   Are these pairs of components *in series*, or *in parallel*?

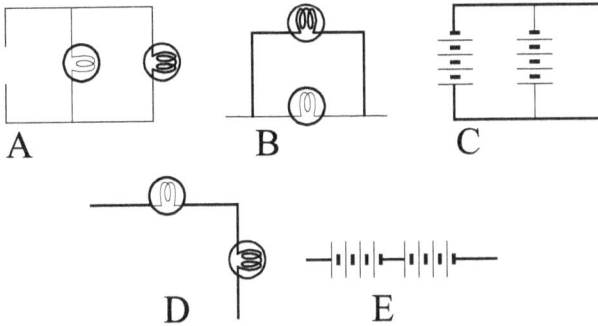

A, B, C, D, E

A_____   D_____

B_____   E_____

C_____

Date: / 4

6   Connect these components with straight lines and square corners, so that they form a complete series circuit.

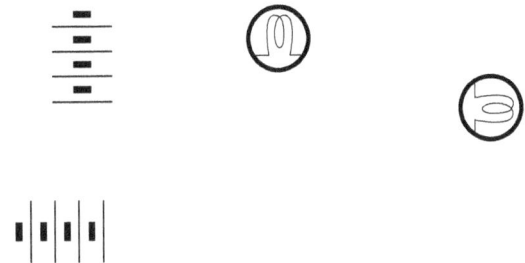

Date: / 4

7   Connect these components with straight lines and square corners, so that they form a complete parallel circuit.

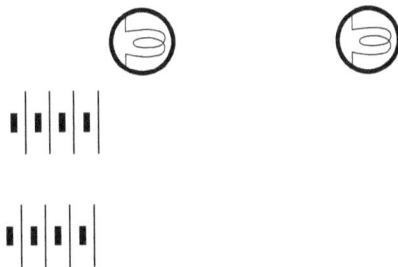

Date: / 4

8   Are these pairs of components *in series, in parallel* or *neither*?

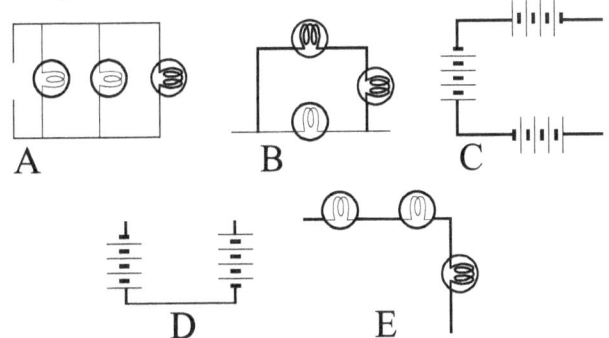

A, B, C, D, E

A_____   D_____

B_____   E_____

C_____

Date: / 4

Quiz 3.4: **Simple Electric Circuits** Name:

---

9 Solve this circuit. (Find V, I, R and P for each component, and totals for the whole circuit). Write all quantities in the spaces beside the letters.

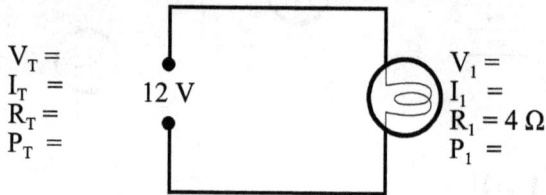

$V_T =$
$I_T =$
$R_T =$
$P_T =$

12 V

$V_1 =$
$I_1 =$
$R_1 = 4\ \Omega$
$P_1 =$

Date: / 4

---

10 Solve this circuit.

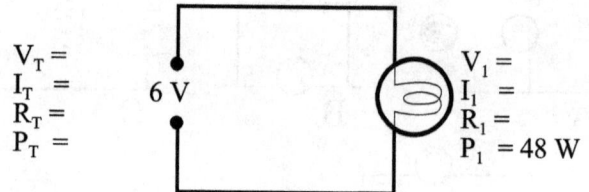

$V_T =$
$I_T =$
$R_T =$
$P_T =$

6 V

$V_1 =$
$I_1 =$
$R_1 =$
$P_1 = 48$ W

Date: / 4

---

11 Solve this circuit.

$V_1 =$
$I_1 = 0.50$ A
$R_1 =$
$P_1 =$

$V_T = 90$ V
$I_T =$
$R_T =$
$P_T =$

Date: / 4

---

12 Solve this circuit.

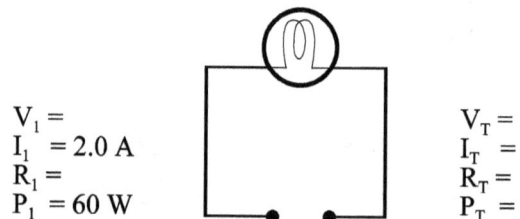

$V_1 =$
$I_1 = 2.0$ A
$R_1 =$
$P_1 = 60$ W

$V_T =$
$I_T =$
$R_T =$
$P_T =$

Date: / 4

All the news that's fit to print... and then some

The Grade Nine Daily

Quiz 3.4: Simple Electric Circuits   Name:

---

13 Solve this circuit.

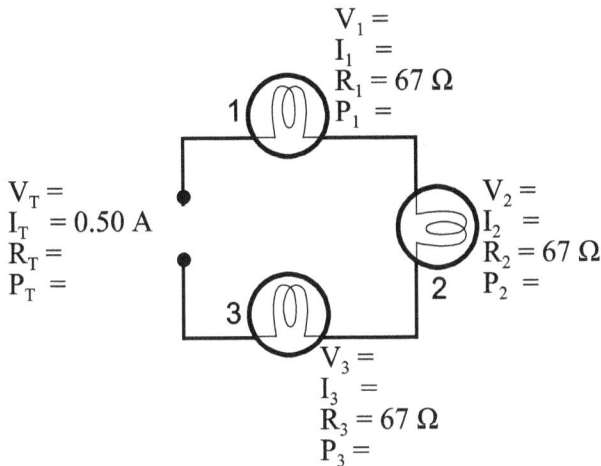

$V_1 =$
$I_1 =$
$R_1 = 67\ \Omega$
$P_1 =$

$V_T =$
$I_T = 0.50$ A
$R_T =$
$P_T =$

$V_2 =$
$I_2 =$
$R_2 = 67\ \Omega$
$P_2 =$

$V_3 =$
$I_3 =$
$R_3 = 67\ \Omega$
$P_3 =$

Date: / 4

---

14 Solve this circuit.

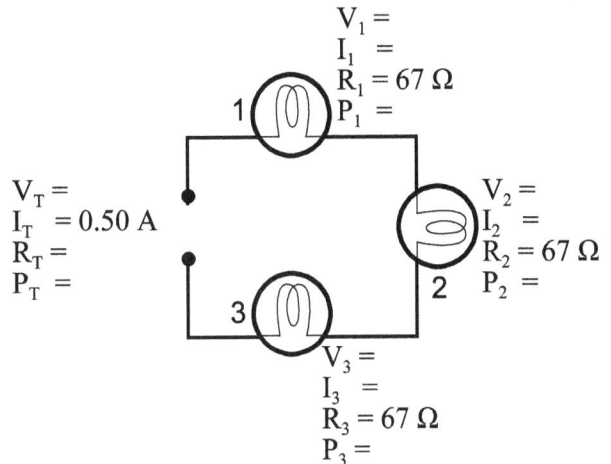

$V_1 =$
$I_1 =$
$R_1 = 67\ \Omega$
$P_1 =$

$V_T =$
$I_T = 0.50$ A
$R_T =$
$P_T =$

$V_2 =$
$I_2 =$
$R_2 = 67\ \Omega$
$P_2 =$

$V_3 =$
$I_3 =$
$R_3 = 67\ \Omega$
$P_3 =$

Date: / 4

---

15 Solve this circuit.

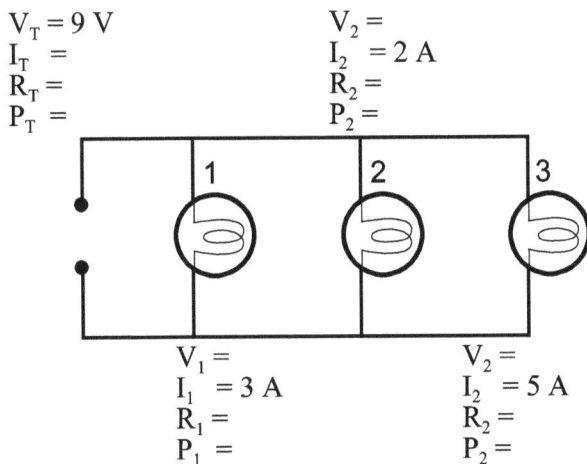

$V_T = 9$ V
$I_T =$
$R_T =$
$P_T =$

$V_2 =$
$I_2 = 2$ A
$R_2 =$
$P_2 =$

$V_1 =$
$I_1 = 3$ A
$R_1 =$
$P_1 =$

$V_2 =$
$I_2 = 5$ A
$R_2 =$
$P_2 =$

Date: / 4

---

16 Solve this circuit.

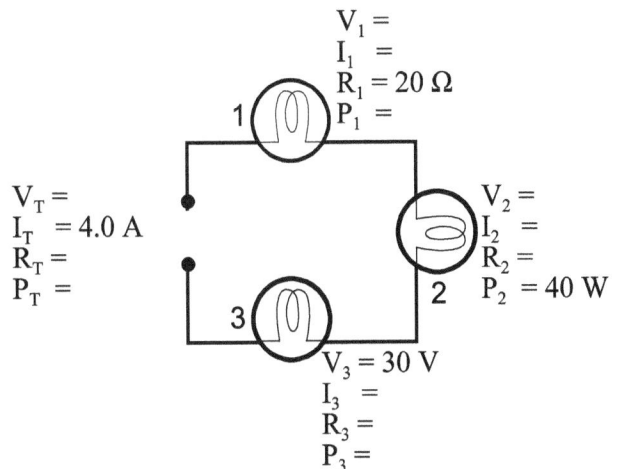

$V_1 =$
$I_1 =$
$R_1 = 20\ \Omega$
$P_1 =$

$V_T =$
$I_T = 4.0$ A
$R_T =$
$P_T =$

$V_2 =$
$I_2 =$
$R_2 =$
$P_2 = 40$ W

$V_3 = 30$ V
$I_3 =$
$R_3 =$
$P_3 =$

Date: / 4

Lab 4.1: Getting Three Light Bulbs to Glow

What's The Question?

Now that you know something about electricity and light bulbs, you will experiment with much more complicated systems. You will be given a battery, wires, and three identical light bulbs.

How many different ways can you connect three identical light bulbs to a source of electric potential so that all three light up at the same time?

What Are We Doing?

1. Draw as many different ways as you can think of to connect three light bulbs together.

2. Predict the brightness of each bulbs. Explain your predictions

3. Experiment. Draw diagrams of your working models, including brightness. Explain each model.

What Are We Thinking About?

1. Two light bulbs 1R and 2R are *in series* if all of the current from 1R must also go through 2R (no branches).

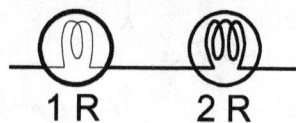

1 R   2 R

2. Two light bulbs 1R and 2R are *in parallel* if both are connected to the same two points (two branches). Current can go through either 1R or 2R, but not both.

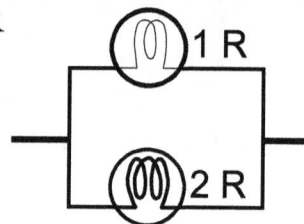

1 R

2 R

Questions For Later...

1. What happens to the total amount of *electric current* at each branching point?

2. Do all of the electrons at any one branching point have the same *potential energy*?

3. Compare two different paths from the first branching point until they come back together at the second branching point. Which electrical quantities are the same, and which are different in the two paths?

Series and Parallel Circuits

Focus Question: Write the question that you are trying to answer.

1 **Predict** as many different arrangements as you can in which three light bulbs light up. Predict the brightness of each bulb. Use a second page if you need to.

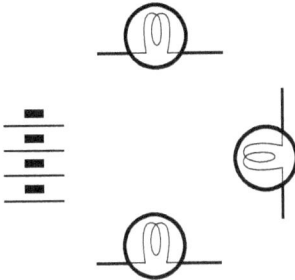

2 **Explain** how your circuit works, in pictures and in words. Use a second page if needed.

3 **Observe** which arrangements work, and how bright the bulbs are. Draw pictures here.

4 **Explain** why your arrangements work the way that they do. You will probably need half a page for each circuit.

One step each day, done by Friday...

The Five Day Project

Project 4.2: Predicting and Measuring Potential

0 Instructions: In Lab 1.11 you learned how to connect three light bulbs in a variety of ways so that they all light up. You will be given three resistors of known value. Connect these resistors so that current flows through every resistor. Predict, and explain, the potential and the current in each resistor. Measure the potential using a voltmeter, to test your predictions. Your teacher will measure the current in each resistor.	**0** My Plan and Outline This column is for your plans, your outlines, and your day-to-day jottings. Plan ahead. Come prepared. For example, on Monday night, write your plan for Tuesday's class. Hand in this sheet along with your four-page report. Date: _____ / 4
1 Obtain three resistors of known value. Label them R_1, R_2, and R_3. Figure out at least four different ways to connect the resistors to give you a truly different circuit each time. Draw and label the circuit diagrams In each circuit, identify all resistors connected *in series*, and resistors connected *in parallel*.	**1** Make records of your circuits. Include all of your attempts, the acceptable ones as well as the rejects. Date: _____ / 4

2 Refer to your text or other sources (not students) to learn how to calculate potential and current for resistors in series and resistors in parallel
 	
Use the equations to calculate:
   V_1, I_1 and P_1 for R_1
   V_2, I_2 and P_2 for R_2
   V_3, I_3 and P_3 for R_3
	2 What equations are appropriate? Cite your sources, using the format suggested by your teacher.
	
	Date: _____ / 4

Project 4.2: Predicting and Measuring Potential   Name:

3 How do you measure potential? Refer to your text or another source to find how a voltmeter works, and how to use it. Refer to your text or another source, and find out how an ammeter works. Redraw all of your circuits, and include three voltmeters and three ammeters in each circuit, along with your predicted values for potential and current.	**3** Write some notes here as you read. Date: / 4
4 Use the voltmeters to measure the potential across each resistor. Ask your teacher to measure the current through each resistor. Redraw all of your circuits, including three resistors, three voltmeters, three ammeters, and the actual measurements for each.	**4** How would a circuit look with voltmeters and ammeters correctly connected to each resistor? Date: / 4
5 Write one full page for each of the following sections:. **Predict:** Your V, I and P values for each R. (from Day 3) **Explain:** your predictions, using diagrams and math. **Observe:** your measurements. Use the circuit diagrams (from Day 4). **Explain:** any differences between your predictions and observations. Attach this page to your report.	**5** Outline your report, especially what you learned at each stage. Date: / 4

Project 4.3: Common Electrical Devices

0   ***Project Instructions*** Choose a small, common electrical device, one that you can actually get your hands on and find out how it works. Some examples are:

Tri-light bulb, fluorescent light fixture

Toaster, kettle, hair blow-dryer, shaver, any other small appliance

Camping lantern, barbecue spark lighter

Controls, such as dimmer switch, fuses, double pole-double throw switches, circuit breakers, thermostats

Do not attempt to plug these devices in!

---

0   My Plan and Outline

Date: / 4

---

1   Choose a device.   Find a physical example that you can work with. Write one full page describing:

Where is it found? Where did you get yours?

What is it supposed to do?

At what voltage does it operate?

What safety issues arise with its use?

---

1   Rough notes and plans here. Write enough that your teacher can understand what to expect in your one page report.

Date: / 4

---

2   Figure out the path of the electricity through the device, and draw two diagrams:

The first diagram is a simplified drawing of the device, as it appears to the eye.

The second diagram is a schematic drawing that only contains the conductors, switches, and resistances.

---

2   Rough notes and plans here. Write enough that your teacher can understand what to expect in your one page report.

Date: / 4

3	Calculate the potential, current, power and resistance in the device when it is operating as it should.	3	Rough notes and plans here. Write enough that your teacher can understand what to expect in your one page report.
			Date: _____ /4

4	Write a step - by - step explanation of how the device works, starting at the point where the high energy electrons enter the device, and ending where the low energy electrons leave the device. This should be about one page.	4	Rough notes and plans here. Write enough that your teacher can understand what to expect in your one page report.
			Date: _____ /4

5	Write a four page report about your device, including: Introduction (Day 1) Diagrams (Day 2) Calculations (Day 3) How it Works (Day 4) Safety and Society issues (Day 5). What issues arise as to excess energy use, safety and environmental concerns with the use of this device? Explain. Attach this page to your report.	5	Rough notes and plans here. Write enough that your teacher can understand what to expect in your one page report.
			Date: _____ /4

The Hazards **In this column is a list of lab safety issues that you will face in this course.**	The Safe Way **Read this column to find out how to safely handle the laboratory problem.**
Eye Injury is possible from flying fragments of metal, glass or chemicals; from heat or flames; from caustic solutions such as acids or bases.	*Always wear safety glasses* in the laboratory. Never take your glasses off, even if you have finished your experiment. Other students may not have finished their lab work. The safety glass symbol indicates exercises in which safety glasses *must* be worn.
Crowding, Pushing and Horseplay increase the likelihood of a serious injury.	*Attend to your work.* Stay at the station you were assigned, so that there is room to work safely. If your teacher finds that your behaviour is a safety hazard, he or she may remove you from the lab. There is no place for behaviours which place others at risk of injury. Not at school, not at home and not at work.
Disorganized and Dirty Working Conditions are a hazard wherever they are found.	*Keep Lab Area Clean.* Clean and put away unused equipment. Tell your teacher about chipped, cracked, damaged or broken equipment. Do not leave anything on the floor, the desktop, the sink, or the cupboards that is not supposed to be there.
Broken Glass happens even to careful scientists.	*Do Not Touch* broken glass with your hands. Tell your teacher. When instructed to do so, use a broom to sweep the glass into a dustpan. Dispose of the broken glass in the special container provided. Do not leave it in the regular wastebasket: it could seriously injure a custodian.
Liquid Spills may consist of water, but they may also contain acids, bases, or toxic chemicals. You may not be able to tell the difference.	*Tell your teacher* about any spills immediately. Do not attempt to clean up without teacher instruction. Only if the teacher decides it's safe, use a cloth or paper towels to soak up excess liquid. Wipe the area clean with a damp cloth. Rinse the cloth frequently in fresh water. Wash your hands afterwards.
Solid Spills may consist of highly reactive chemicals. You may not know the specific hazards.	*Tell Your Teacher* about the spill, whether or not you caused it. Your teacher will instruct you on the safe way to handle the problem. In any case, the spill must be cleaned up promptly.

Appendix: Laboratory Safety

Open Flames are a frequent hazard. The Bunsen burner is the most likely safety hazard.	**Review Safe Handling of a Bunsen Burner** with your teacher. Be prepared to show how to light, operate and extinguish the burner at any time. Do not attempt to ignite pens, papers, rulers or other things. That kind of behaviour will certainly result in your being put out of the lab.
Fire. Any liquid solid or gaseous fuel burning where you do not want it to burn is a fire.	**Tell the teacher immediately!** Do not attempt to extinguish the fire with your hands, books, paper towels etc. Do not panic. Move away from the hazard. **Your teacher is the best judge of the appropriate course of action.**
Hot Metal or Glass cause more burns than any other hazard. There is usually no visible indication that they are hot. Glass in particular causes small, deep burns.	**Let Hot Objects Cool for 10 - 15 Minutes** before handling. Place all hot objects on a heat resistant pad. You and your partner will know where they are. Approach hot objects cautiously. Touch them at the coolest point first (the base of the retort rod, the bottom of the Bunsen burner or hot plate, the thumb screw of the iron ring). Use dry, not damp, paper towels to handle hot objects.
Hot Liquids such as boiling water or hot oil spread and splash rapidly. They also cling to skin and clothes.	**Let Hot Liquids Cool for 10 - 15 Minutes** before handling. Do not heat liquids in closed containers. Use hot plates rather than shaky retort rod assemblies. Do not heat more liquid than you need.
Obstructed Passageways prevent you from moving out the way of a spill or a fire.	**Stand at Your Lab Station.** Do not bring chairs or stools over to sit down. Your chair will prevent others from moving away from a spill or a fire.
Long Hair or Loose Clothing is more likely to become involved in your equipment. It can cause spills and breakage, or catch fire.	**Tie Back Long Hair; Secure Loose Clothing.** Outerwear in particular must be avoided in the lab situation. Jackets, sweat suits, hoods, etc are too large and awkward for the lab situation. They are also frequently made of materials that are flammable and can melt and stick to the skin in a fire. Avoid using laquer based hair sprays. A curly head of hair with hair spray can burn up completely in seconds.
Unauthorized Experiments can have unintended results.	**Stick to the plan.** Read instructions very carefully the night before the lab. Ask questions. Do not try experiments "just to see what happens." The dangers are too great.

www.ingramcontent.com/pod-product-compliance
Lightning Source LLC
Chambersburg PA
CBHW051229200326
41519CB00025B/7306